More Praise for *THE BRIGHT SIDE*

"A WHOLE NEW LENS THAT PROMISES TO TRANSFORM HOW WE NAVIGATE THE TWENTY-FIRST CENTURY. Many books on optimism take an individualist approach; *The Bright Side* is altogether more ambitious: this is a guide for civilization . . . wise about the lessons of the past, clear-sighted about the challenges of the present, and hopeful about the possibilities of tomorrow."

—Richard Fisher, author of *The Long View: Why We Need to Transform How the World Sees Time*

"A CHARMING AND GENEROUS INVITATION TO THE FUTURE . . . *The Bright Side* is a still, small voice of calm in a culture hijacked by apocalyptic thinking and regimented behavior."

—Simon Ings, author of *The Weight of Numbers*

"A TONIC FOR OUR TIMES. Paul-Choudhury showed impressive courage after his young wife's death, vowing to map the intellectual and psychological underpinnings of optimism. We are the beneficiaries of his tenacity. In lucid, lively, never highfalutin prose he brings onstage such characters as Odysseus, Oedipus, Gottfried Wilhelm Leibniz, and Helen Keller, each of whom has a vital lesson to impart. The book's conclusion: not only is optimism rational, it's a twenty-first century necessity."

—Meredith Wadman, author of *The Vaccine Race: Science, Politics, and the Human Costs of Defeating Disease*

"A SALVIFIC, JOYOUS PAEAN TO ALL THAT'S POSSIBLE. Read this book. Don't be the version of yourself that doesn't: because the best possible version of you is almost certainly the one that picks it up. Peppered with the effortless wit—and profound curiosity—that only a true optimist can muster, Paul-Choudhury's *The Bright Side* is a panoramic guide to why we should always hope that things can get better, even when it seems most likely they will not. It is just the book this world needs."

—Thomas Moynihan, author of *X-Risk: How Humanity Discovered Its Own Extinction*

"MOVING AND INSIGHTFUL . . . This argument for optimism stands up against the cynical mood of the twenty-first century, challenging—and equipping—the reader to help create a better world."

—Michael Brooks, author of the bestselling *13 Things That Don't Make Sense*

"A SPELLBINDING TOUR DE FORCE. Paul-Choudhury has crafted an expansive and convincing argument for why building a better future demands that we believe in it first. An urgent call to arms to choosing change over doom, action over despair, optimism over hope."

—Roberto Trotta, astrophysicist and author of *Starborn*

"TIMELY . . . This is a book about the future for anyone who believes there isn't one, a reminder that instead of worrying about what we expect to happen, we should focus our energies on whatever it is that we, collectively, want to happen."

—Richard Watson, author of *Digital vs. Human: How We'll Live, Love, and Think in the Future*

"DEFTLY MELDS SCIENCE, PHILOSOPHY, AND CULTURE to build a convincing case for optimism. I left this book feeling energized and empowered."

—David Robson, author of *The Expectation Effect: How Your Mindset Can Change Your World*

The Bright Side

How Optimists Change the World, and How You Can Be One

Sumit Paul-Choudhury

SCRIBNER
New York Amsterdam/Antwerp Toronto London Sydney New Delhi

For a list of sources for this book, please visit alternity.com

Scribner
An Imprint of Simon & Schuster, LLC
1230 Avenue of the Americas
New York, NY 10020

Copyright © 2025 by Sumit Paul-Choudhury

All rights reserved, including the right to reproduce this book or portions thereof in any form whatsoever. For information, address Scribner Subsidiary Rights Department, 1230 Avenue of the Americas, New York, NY 10020.

First Scribner hardcover edition January 2025

SCRIBNER and design are trademarks of Simon & Schuster, LLC

For information about special discounts for bulk purchases, please contact Simon & Schuster Special Sales at 1-866-506-1949 or business@simonandschuster.com.

The Simon & Schuster Speakers Bureau can bring authors to your live event. For more information or to book an event, contact the Simon & Schuster Speakers Bureau at 1-866-248-3049 or visit our website at www.simonspeakers.com.

Manufactured in the United States of America

1 3 5 7 9 10 8 6 4 2

Library of Congress Cataloging-in-Publication Data has been applied for.

ISBN 978-1-6680-3140-7
ISBN 978-1-6680-3142-1 (ebook)

To my parents,
who gave me the best of all possible worlds
—and my children,
to whom I owe the best of all possible futures.

Contents

PROLOGUE

The Case for Optimism

Why be an optimist?

I became an optimist the night my wife died.

People reacted in different ways to her diagnosis: an aggressive, fast-spreading ovarian cancer discovered after the miscarriage that ended our first and only pregnancy. A few understood that her future was likely to be grim and short; those people mostly kept quiet or stayed away. But many professed to believe that things would somehow work out—sometimes out of superstition, sometimes out of desire to reassure, but most often simply because they could think of no other way to react.

Kathryn, for her part, insisted that those around her—her family, her friends, her colleagues and her doctors—only express hope. Naturally, that applied to me most of all, but I struggled to know how to accommodate her wishes. On the one hand, I'd always been inclined to look on the bright side, and some part of me believed it would all work out fine. On the other, I was an empirically minded rationalist. I read the medical reports and the scientific literature, and realised that her odds of surviving more than a couple of years were vanishingly small. But since I wasn't the one with the terminal illness, I concluded that I should keep my mouth shut and be supportive in the way my wife had chosen, while hoping against hope for a statistical miracle.

No miracle came. Kathryn's cancer overran her body's defences in less than a year, and she endured an unrehearsed and graceless death.

When it came to rebuilding my own life, the piece of advice I was given, over and over, was to "take it one day at a time." No long-term plans, no significant life changes. I found that unsatisfactory. While there were clearly some decisions it would have been unwise to make while still bowed over in grief, I didn't want to spend any more time in limbo than I already had. It helped that Kathryn had told me, in no uncertain terms, that I wasn't to lose my way once she was gone—however difficult, she wanted me to keep moving forwards.

Most of us, most of the time, walk the path of least resistance. I certainly had: though I didn't particularly like where I lived or what I did for a living, it was comfortable enough. But after Kathryn died, my home was no longer my home, and my future no longer my future. I certainly don't *recommend* bereavement as a way of hitting the reset button, but it did give me the opportunity and motivation to rethink my life from scratch. It forced me to consider all the possible ways in which I might reconstruct it. At least I still *had* possibilities to explore.

I started trying on different lives for size: the rural hermit, the man-about-town, the perpetual nomad. A lifelong urbanite, I spent much time tramping in wilds and woods. Without reasons to stay home, I spent my days at galleries and nights at gigs, pursuing moments of escape that rarely came. I'd always travelled as much as I could, but I started crossing destinations off my bucket list in short order. And I started two daily blogs, one for friends and one for the world, to write myself into the future: mournings, imaginings and beginnings.

Some months in, a well-meaning friend asked if I was still on medication. It had never occurred to me to take any. Ditto counselling: I was gently nudged towards it but gently nudged myself away again. I found a support group for young widowers, but I'd never been the kind of person who'd join any club that would accept me. Many in my position observed birthdays, anniversaries, holidays; after the first year, I made the difficult decision to stop observing them. I didn't want the rest of my life to run to an out-of-date schedule.

It gradually dawned on me that my approach wasn't entirely typical. I wondered if I was in denial. Or perhaps just an emotionless brute. It didn't *feel* as though either of those was true: I was by no means happy or normal during my period of mourning, I just never doubted, even on my darkest days, that better times lay ahead—if I only worked towards them. Initially, without really thinking about it, and later more deliberately, I cultivated the idea that the future would be bright. Eventually, I realised that I'd chosen to identify as an optimist.

That was somewhat perplexing. As a trained scientist and a working journalist, I was supposedly a hardened critical thinker, committed to solid evidence and rational argument. While I knew, and had been told, that I tended to expect the best out of life, I'd presumed that was because I actually *had* led a pretty charmed life. To *still* expect that, after the events of the previous year, felt as if I had given myself over to irrationality: the side of me that *believed* was winning out over the side that reasoned.

My impression of optimism was that it amounted to nothing more than a belief, and that to place any weight on it was fundamentally silly and potentially irresponsible. Calling yourself an optimist seemed like admitting that you just didn't want to think very hard about the future and its challenges. But at the same time, what was the alternative? The usual defence of pessimism is that a pessimist is never disappointed, and can only ever be pleasantly surprised. That seemed a needlessly defensive, almost cowardly stance. And professed "realism" seemed to me to be fence-sitting, a cynical excuse to avoid engaging with the possibility that the world could be better than it is today.

I couldn't see how either of those worldviews would propel you through life. Why would you even bother to get up in the morning?

Once I started to think about it, optimism seemed like the only stance worth taking. At least expecting more out of life primed you to *get* more out of life, or so I assumed. But if I was going to be an optimist—and I appeared to have little choice in the matter—then I wanted to practise a kind of optimism for which I could articulate a defence that amounted to more than just belief. I wanted to find a way of being an optimist that might actually help make the world better, rather than just assuming it somehow would be.

So I began to investigate what form that pragmatic, well-reasoned version of optimism might actually take. And what I learned was that optimism, despite my earlier assumptions, isn't necessarily the product of naivety. It isn't an indulgence that we can only afford when times are good. It's a resource we can tap into when the going gets tough—and then it can make the difference between life and death.

Asses and Lions

"It was a sickening sensation to feel the decks breaking up under one's feet," recounted Ernest Shackleton, "the great beams bending and then snapping with a noise like heavy gunfire." After more than nine months stuck fast in the Antarctic ice, his expedition's flagship, the *Endurance*, had finally succumbed. And so, on 27 October 1915, the great Anglo-Irish explorer reluctantly ordered his crew to abandon ship—leaving them stranded in perhaps the planet's most hostile location.

Shackleton's Imperial Trans-Antarctic Expedition had set off more than a year earlier with the objective of making the first land crossing of the frozen southern continent. But first it had to traverse the remote Weddell Sea, an extraordinarily difficult area to navigate. Its progress through the pack ice was slow, and stopped altogether in January 1915, forcing Shackleton and his men to survive the bitterly cold, permanently dark Antarctic winter—conditions that would still be supremely challenging today.

Once the *Endurance* had sunk, Shackleton and his men had to make do with the equipment and provisions they had crammed into its three lifeboats. Other than that, they had to live off their wits and the scant resources the terrain provided. (On one occasion, they killed an aggressive leopard seal; its stomach contents of as-yet-undigested fish provided a welcome "fresh" meal.) They had no way of communicating with the rest of the world and essentially no hope that anyone would come to rescue them. And yet, ten months later, Shackleton led every one of his men to safety.*

How did Shackleton and his crew pull through in the face of such incredible odds? The practical answer is that they spent more than five months camped out on the ice before setting sail in the lifeboats to Elephant Island, which was solid land but had little else going for it. From there, Shackleton took one of the boats and a skeleton crew on a two-week, 800-mile journey through raging seas to South Georgia. On 20 May 1916, he finally reached a whaling station on its north coast, whereupon he promptly borrowed a ship to mount a rescue mission. That rescue was frustrated by impassable ice, as were three further attempts. But in August he finally retrieved the twenty-two men still waiting on Elephant Island.

Behind this practicality lay persistence. As countless leadership texts point out, this incredible derring-do would never have been possible had Shackleton and his crew not shared a common

* Every man on the *Endurance* survived, but the receiving party on the other side of Antarctica fared badly: unaware of Shackleton's difficulties, they went to extraordinary but ultimately pointless lengths to lay out supply dumps for his anticipated crossing. Several men died in the process.

bond and strength of purpose: many previous expeditions had come to grisly ends when their team members fell out with each other after encountering difficulties. Everything that Shackleton achieved, he achieved with and because of those he had taken with him—twenty-seven men chosen from more than five thousand applicants. What did he look for?

The usual story is that he had recruited adventurers through an advert in *The Times*, stating: "Men wanted for hazardous journey. Low wages, bitter cold, long hours of complete darkness. Safe return doubtful. Honour and recognition in event of success." But there's no record of such an ad ever appearing. And actually, Shackleton wasn't looking for tough guys. His criteria could appear quixotic, including everything from physical appearance to sense of humour, but, he said, "the quality I look for most is optimism: especially optimism in the face of reverses and apparent defeat. Optimism is true moral courage."

What did Shackleton mean by optimism? It certainly wasn't blind faith that things would work themselves out. Shackleton was known for carefully evaluating his options at every turn: his men called him "Old Cautious." On an earlier voyage, the Nimrod Expedition of 1908–09, he had turned back less than a hundred miles from the South Pole after concluding his party wouldn't be able to make the return trip. "I thought, dear, that you would rather have a live ass than a dead lion," he wrote to his wife.

At every stage of the troubled Imperial Trans-Antarctic Expedition, Shackleton reminded his men that as long as they were alive, they had choices to make and options to explore, inspiring his weather-beaten, malnourished and profoundly isolated crew to keep the faith. He also needed them to share in that spirit to fend off potentially fatal disputes, which was especially true for those left on Elephant Island under the leadership of his second-in-command, Frank Wild. "His cheery optimism never failed, even when food was very short and the prospect of relief seemed remote," wrote Shackleton. "I think without doubt that all the party who were stranded on Elephant Island owe their lives to him. The demons of depression could find no foothold when he was around."

Not many of us will have our mettle tested as Shackleton and his team did. But we all have our reckonings with life and death sooner or later, or other adversities that make us reappraise the world and question the future. It's at such times that optimism can be hardest to secure, but also most valuable. Its value is also evident when it comes to more mundane challenges: a date, a job interview, a sports match. If we are downcast about some situation or other, the chances are that someone will tell us to look on the bright side; and the chances are that we will feel better if we do. And we may well find a solution to our problem once our mood lifts, if there is a solution to be had.

Optimism, far from leading us to passively await our fates, can help us to actively explore our limitations—and transcend them.

Within and Without

"Once I knew only darkness and stillness. Now I know hope and joy," wrote Helen Keller in her 1903 essay "Optimism," twenty-one years after the childhood illness that had destroyed her sight and hearing. "My life was without past or future; death, the pessimist would say, 'a consummation devoutly to be wished.' But a little word from the fingers of another fell into my hand that clutched at emptiness, and my heart leaped to the rapture of living."

The fingers were those of Anne Sullivan, Keller's teacher; and as generations have learned, the "little word" they were spelling out was "water," which was simultaneously running onto Keller's other hand. Sullivan had previously worked with Samuel Gridley Howe, the doctor who had pioneered the finger-writing technique two decades earlier with Laura Bridgman, another young deaf-blind girl. Keller's mother had read Charles Dickens's touching account of their success, and sought help for her own frustrated daughter, then aged seven.

In her essay, Keller suggests that Howe "found his way to Laura Bridgman's soul because he began with the belief that he could reach it. English jurists had said that the deaf-blind were idiots in the eyes of the law. Behold what the optimist does. He converts a hard legal axiom; he looks behind the dull impassive clay and sees a human soul in bondage, and quietly, resolutely sets about its deliverance." Keller, thus delivered, went on to become an author, speaker and activist—achievements so extraordinary that many believed, by way of a backhanded compliment, that she must have been faking her condition.*

The power of optimism became a recurring theme in Keller's work. Her praise for it did not stop at what she referred to as "Optimism Within," the internal, personal conviction that better times are to come. This "fact within my own heart," Keller wrote in the second part of her 1903 essay, was mirrored by "Optimism Without," the belief that the condition of the world has steadily improved—materially, socially and spiritually. This march of progress, she suggests, is evident from "literature, philosophy, religion and history."

The third part of her essay, "The Practice of Optimism," opens by noting that "the test of all beliefs is their practical effect in life." Optimism, Keller suggests, "compels the world forward, and pessimism retards it." Pessimism for a nation, as for an individual, "kills the instinct that urges men to struggle against poverty, ignorance and crime, and dries up all the fountains of joy in the world." Optimism, by contrast, is "the faith that leads to achievement." Without it, nothing can be made better.

Keller's description of optimism's power, effects and practice still resonate today. Her God does not promise a better world by

* This scepticism endures: in recent years "Helen Keller denialism" has circulated among TikTokers who don't believe that what they're hearing in school is a true story.

fiat; instead, the glad thoughts He inspires must lead to practical actions. Her America may be the finest nation the world has yet produced, but she cautions against the "dangerous optimism of ignorance and indifference." An optimist may believe the world to be basically a good and just place, but must nonetheless pledge to recognise and resolve whatever suffering exists.

It's of its time. Our worldview has been transformed over the past century: where Keller had faith, patriotism and destiny, we have psychology, philosophy and prediction. But her concerns are echoed in the pages of this book, and that's why, 120 years later, I've chosen the same structure for it as she did for her essay. In the first section, "Optimism Within," we'll look at what we know about our inner, intuitive optimism, and what it does for us. In the second part, "Optimism Without," we'll ask if there's any rational, intellectual basis for expecting the best from the world. And in the third, "Optimism in the World," we'll look at our power to anticipate what lies ahead—and change it for the better.

Traps and Gaps

For all this talk of the power of optimism, there's a problem. Optimism is associated with unexpected victories, mountains moved and triumph over adversity; but also with unkept promises, unaffordable bets and unrealised dreams. Our conception of optimism is about positive expectations of the future—although it hasn't always been, as we'll see—and of course we can never *prove* that those expectations are well-founded before the fact. When things turn out well, we praise the optimism of our political, social and commercial leaders as inspirational; when things go badly, we disparage it as wishful thinking.

Lots of people have tried to patch things up by explaining how to modify, or perhaps tame, optimism to make it more disciplined. Blog posts, think-pieces, interviews and explainers offer abundant variations on the theme: "conditional optimism," "indefinite optimism," "optimistic nihilism," "tragic optimism," "cruel optimism," "pragmatic optimism," "rational optimism," "nauseous optimism," "epistemological optimism," "apocalyptic optimism" and so on ad infinitum.

Some of these formulations have their uses, but the overall effect is of cringing diffidence. Better to accept that optimism is unrealistic but works for us nonetheless. After all, what does "unrealistic" mean anyway? We could flip things around and ask how many supposedly "realistic" expectations of the world actually *are* realistic, given the profound uncertainties we face. If we were to apply the same sort of linguistic sleight of hand, we'd find ourselves talking about "spurious" or "self-interested" realism, "fatalistic" or "lazy" realism, or just plain "bullshit realism."

We often just don't understand enough to be "realistic" about the challenges we face, like climate change, societal upheaval or artificial intelligence. More often, there is a range of possible outcomes—degrees of warming, shades of civil disorder, uses and abuses of technology—over which we have varying degrees of control. That's not to say we don't know anything at all: we have developed many sophisticated ways to assess our present situation and future scenarios. We can and should continue to improve our ability to model and predict what's coming. But we should also accept that we don't and can't know everything: the future is elusive.

That, in fact, has always been the human condition, and we have the cognitive tools to deal with it. Our brains are the product of millions of years of evolution. That doesn't mean all the ways our minds work are perfectly suited to the challenges of modernity, but it would be unwise to disregard our intuitions and imaginations, as well as our intellects. If we want the world to be better tomorrow than it is today, we first have to *expect* that it will be. Then we have to *imagine* the ways in which it could be. And then we have to *ensure* that it will be.

So one way to make the case for optimism is to acknowledge that there are things we don't know, that some of those unknowns are positive, and that we have some ability to steer towards those positives. Optimism encourages us to seek them out. If, on the other hand, we have no expectation that our lot in life can be improved, we have no motivation to put in the thought and effort needed to improve it and those solutions go undiscovered. Failure becomes a self-fulfilling prophecy.

There's a parable that illustrates this neatly. Nineteenth-century travellers noted that Russian peasants had a novel way of keeping their milk fresh in the absence of refrigeration: they would put a frog in it.* Whether or not this ever actually happened, a folktale describes the fate of two such frogs placed in neighbouring churns. One, having established that it's impossible to climb or jump out, despairs of its fate and soon drowns. Its more determined companion, however, continues to kick and thrash—until eventually the milk is churned into butter, whereupon it leaps to freedom.

The first frog has fallen into a "pessimism trap": because it doesn't see any way out of its predicament, it simply gives up, and its demise becomes a self-fulfilling prophecy. The second frog, on the other hand, continues to struggle even though it can't see any way to escape either, and in so doing, stumbles on a solution it could not have predicted at the outset. That second frog is an optimist.

Pessimism traps abound in human lives. Jobs you don't expect to get, so you never apply; crushes whom you believe to be out of your league, so you never ask them out; games you expect to lose, so you never play. From this point of view, it's not surprising that optimists turn out to be more successful than pessimists in almost every aspect of their lives. They tend to do better at school and at work; they have stronger relationships with family and friends;

* They might just have had a point. In 2012, scientists from Russia (where else?) found that some frogs secrete a remarkably large number of antibacterial chemicals through their skins.

and they're more resilient in the face of financial, mental or physical stress. It's the stuff of cliché. Eighty per cent of success is just showing up. You miss every shot you don't take.

While we're well aware of pessimism traps in our own lives, we often find it harder to avoid them in the wider world. As the philosopher Jennifer Morton, who popularized the concept, observes, surveys show that Americans report far higher confidence about their own prospects than their country's, and increasing levels of gloom about the future—particularly among the young people who will have to live there. In Europe, too, a survey of 12,000 citizens in December 2019 found that 58 per cent felt positively about their individual futures—but only 42 per cent felt that way about the future of their countries.* Interestingly, the pattern is less evident in poorer countries, but in the wealthy ones there's frequently a gap between our personal and social expectations—an "optimism gap," as the writer David Whitman called it.

To be sure, there are serious challenges and tough times ahead. But when it comes to, say, climate change, we're actually better off than the frogs, because we know what the answer is: eliminating carbon emissions. And we know how to achieve it: solutions have been proposed for every imaginable part of the problem. But if we don't expect that we can bring about change, we won't act in ways that bring change about—which leaves us meekly accepting whatever the future may hold, and there's no reason to expect that future will be positive. On the contrary, it's likely to be dire. The optimism gap will have mutated into a pessimism trap. Just like the first frog in the milk churn, we will have given up on getting out.

Optimism gaps and pessimism traps pose a real threat to our ability to solve the genuine and urgent problems of our age. They sap us of the motivation needed to seek out solutions, and of the

* Many surveys gauge life satisfaction, well-being or happiness. Optimism is related to those qualities, as we'll see, but it isn't identical to any of them; and fewer surveys ask about it.

will to put them into practice. The young need reassurance that the future is still within their grasp; their elders need to fight back against the inertia that propels us into gloom. Finding those solutions will take ingenuity and effort—and the willingness to look for them—even if we don't know at the outset what they are, or where to find them.

On 24 May 1963, the American writer and civil rights activist James Baldwin was interviewed for television by his fellow activist, the psychologist Kenneth Clark. It was meant to be a wide-ranging discussion of the civil rights movement, following a tumultuous few months during which protests in Birmingham, Alabama—reputedly the most segregated city in America—had been put down with fire hoses, police dogs and mass arrests. One of those arrested was Martin Luther King Jr., whose famous "Letter from Birmingham Jail" made the case for civil disobedience in the face of unjust laws.

But earlier that same day Baldwin, Clark and other activists had been invited to meet Robert Kennedy, then the US attorney general. The meeting had become acrimonious, leaving both sides frustrated and sceptical that progress could be made. The experience seems to be weighing on Baldwin and Clark in the televised interview: both are subdued, troubled. Cigarette smoke coils around them. Baldwin answers Clark's first question, about his experience of education, but abruptly pivots to the brutality meted out to the Birmingham protestors, in particular an incident in which five policemen pinned down a woman, one putting his knee on her neck. Eventually Clark asks: "Jim, what do you see deep in the recesses of your own mind as the future of our nation? . . . What do you see. Are you essentially optimistic? Or pessimistic?"

"Well I'm both glad and sorry you asked me that question. And I'll do my best to answer it," replies Baldwin. "I can't be a pessimist," he says with a shrug. "Because I'm alive. To be a pessimist means that you have agreed that human life is an academic matter. So I'm forced to be an optimist. I am forced to believe that we can survive whatever we must survive."

It can't have been evident to Baldwin and Clark in that TV studio, but 1963 was to prove a pivotal year for the US civil rights movement. After their meeting, Kennedy ordered the FBI to intensify its surveillance of Baldwin and other activists, but also became more sympathetic to their cause. Baldwin published *The Fire Next Time*, which found a receptive audience for its message that activists "may be able, handful that we are, to end the racial nightmare, and achieve our country, and change the history of the world." In August, Martin Luther King Jr. gave his "I Have a Dream" speech, one of the most optimistic pieces of oratory ever delivered. The struggle for civil rights continued—King's speech was swiftly followed by brutal violence from white supremacists—but a way forward had opened up where none had been visible before.

Baldwin was right: human life is not an academic matter. The answers to what we can expect from our lives, what kind of world we live in, what we can do to shape the future—for all that these questions have been debated for millennia—lie not in reason, but in conviction, determination and belief. We live on the verge of momentous change, perhaps of crisis. As another political activist, the Italian Antonio Gramsci, put it in a letter from one of Mussolini's jails: "The old world is dying, and the new one struggles to be born; now is the time of monsters."

Now, as then, an old world is dying. Our job is to help that new world to be born, and ensure it is not midwifed by monsters. Not a perfect world, but a better world, one born not only from the academic virtues of reason, intellect and planning, but also from the irrational ones of belief, imagination and possibility: from optimism.

PART I

Optimism Within: Positive Illusions

Why are we naturally optimistic?

Is the world sliding into a pessimism trap?

Can we learn to look on the bright side?

1

Optimistic Apes

Why are we naturally optimistic?

Why did the chicken cross the road? Because it believed it could get to the other side.

It's hard to tell if a chicken is an optimist. You can't ask it if a glass of water is half full or half empty, after all. But what you *can* do is to repeatedly show a chick a white card in front of a bowl of tasty mealworms and a black card in front of an empty bowl. Once the chicks have learned to reliably choose the white card, you show them a *grey* card. Chicks that head immediately for the grey card are considered optimists. The card is more white than black, they apparently surmise—the equivalent of deeming a glass half full rather than half empty. And on this basis, most chickens do indeed turn out to be optimistic.

Chickens are far from the only animals to show this sort of "judgment bias." The first such tests were conducted in the early 2000s, on rats who had learned that pressing a lever when they heard one tone resulted in delivery of a food pellet, whereas pressing it in response to another tone produced only an unpleasant burst of white noise. Since then, scientists have adapted this kind of "ambiguous stimulus" test for many different species, and have tried many variations to explore the influence of different environments and experiences. And the more experiments they do, the longer and more esoteric their list of findings become: Left-pawed dogs are more "pessimistic" than those which prefer their right paw. European starlings become more "optimistic" if they can take a bath whenever they want. Cows learn faster when puffed with air than when given an electric shock. Mini pigs pick up the test more quickly than farm pigs, but that doesn't make them any more positive. Sheep given stress-inducing drugs don't become more pessimistic. Bottlenose dolphins are more optimistic if they have been swimming in synchrony with each other. And a sip of sugar makes the world seem sweeter to bumblebees.

This might seem like a miscellany of offbeat findings, but these studies are seriously intended. They're looking for clues as to how optimism relates to the well-being of the animals being tested, which is why they're so varied: different species have different requirements for a life well lived. The Swedish team that carried out the chicken study, for example, wanted to know if birds that grew up in stimulating surroundings remained more positive when stressed than those raised in more sterile conditions, with obvious implications for their welfare in agricultural contexts. (They did.)* The net conclusion is that an animal's living conditions shape its propensity to take the "optimistic" route. Stressed animals—ants given mild electric shocks, rats living in continually changing habitats, dogs in the company of stressed *humans*—tend to be less optimistic, whereas those living in enriched surroundings, or who have recently enjoyed an unexpected treat, tended to look (or lick, or sniff) on the bright side.

* Bizarrely, an earlier attempt to improve chickens' well-being resulted in millions of them being fitted with tiny, rose-coloured glasses. It's not as cute as it might sound: the glasses were to protect their eyes if they fought, as they often did when kept in close quarters; the colouration supposedly prevented them becoming frenzied by the sight of spilt blood.

Is this really optimism as humans understand it? As the list above suggests, animals act according to their own, highly species-specific motivations, and it's hard to compare those to our own. But since humans, too, are animals, it seems reasonable that our optimism is another variation on the theme, perhaps reflecting our own requirements for well-being.

If so, you would expect to see it among our closer evolutionary relatives. And sure enough, capuchin monkeys, marmosets and macaques all show judgment biases when given ambiguous-stimulus tests. There's less evidence when it comes to our very closest relatives, the great apes, because these trials are ethically and practically complicated, and thus tend to be very limited. But there are some suggestive, if tentative, findings nonetheless. For example, one trial involving three chimpanzees suggested that their optimism related to, and was enforced by, their strongly patriarchal social ranking: Nicky, the dominant male, was the most positive, and ET, the sole female, was the most pessimistic.

So optimism of a basic but recognisable kind seems to be widespread through the animal kingdom, and it's strongly associated with well-being. But why? How is it better to err on the positive? Wouldn't it be better for an animal—or a person—to make the most accurate assessment possible of their situation, then react accordingly? Perhaps, but the problem is that it's not always, or even usually, possible to make a very accurate assessment.

Imagine you're a mouse. Like most small mammals, you burn energy quickly, so you need to eat often. You'll starve to death if you go more than a couple of days without a meal. The problem is that you burn even more energy while searching for food. So if you leave your cosy burrow in search of seeds, but don't find enough, you'll return home worse off than you were to begin with. How do you know when you should go out foraging? If you go out at every possible opportunity, you'll waste a lot of effort. You might even lose your life. On the other hand, you can't just stay nestled up all the time: if you wait too long to venture out, you'll eventually become too weak to forage at all. So: Should you go out at every opportunity and risk exhausting yourself, or stay home for ever and face the prospect of inevitable starvation?

Obviously the best answer lies somewhere between these two extremes. But where exactly?

In an ideal world, you would be able to judge the situation precisely, knowing everything you needed to about the ripeness of the grain, the lie of the land, the habits of the neighbourhood cat, and so on. You would weigh the odds of success against your growing hunger. If your odds of gathering more calories than you expend are better than fifty-fifty, go out and eat. If they're less than that, bide your time at home.

In the real world, nothing is so clear-cut. You may not know how much food is still available (maybe other mice have beaten you to it), whether the fertile patch you found last time is still intact, or if the cat's route around its territory has changed. So should you stay or should you go? It's a judgment call.* This is the dilemma that's simplified in experiments down to the ambiguous stimulus—the chickens' grey card.

Given all these uncertainties, you're bound to make the wrong decision sometimes. But you need to get it right more than you get it wrong. If you're half-starved—whether because you're forever wasting energy on fruitless foraging runs, or because you go out too infrequently to keep yourself properly fed—you're not going to be successful at attracting a mate, providing for your offspring or out-running predators. Wild recklessness and excessive timidity both result in extinction. Over millions of generations, natural selection will force you to become very good at deciding whether to go out or not.

* This is you thinking as if you were in a mouse's situation. How an actual mouse might gather and process this information is a subject for another book.

The premise of "error management theory," as set out by the behavioural psychologists Martie Haselton and Daniel Nettle, is that a species will become expert at managing those forms of error that are likely to end up with it losing out. It can thus work out better to take action even when your best estimate of success is fairly low, because if your estimate is overly cautious but you act anyway, you'll benefit from opportunities that you hadn't even realised existed. So it can be disproportionately beneficial to favour action when the costs of failing are low (a bit of wasted effort spent foraging) compared to the benefits of succeeding (lots of delicious calories to consume).

That puts some basic logic behind "optimistic" behaviour. Does it translate to humans who, with all due respect to mice, have to navigate rather more complex situations? Perhaps, at its root, our optimism serves to encourage us towards action when the situation defies rational evaluation. Under those circumstances, erring towards the positive might reap unanticipated rewards. We would expect optimists to actively take chances that their more passive cousins forgo, seizing the day and securing advantage. That certainly *seems* to be how optimism works out for people—albeit that our optimism takes its own, distinctively human form.

Positive Illusions

One of the difficulties in understanding optimism is that the word itself is used to mean many somewhat different things. So let's start with the dictionary definition. Merriam-Webster says optimism is "an inclination to put the most favorable construction upon actions and events or to anticipate the best possible outcome." That this statement *sounds* straightforward testifies to how sophisticated our brains are, because it actually implies remarkable capacities for reasoning, imagination and planning.

"Construction" requires us to imagine a scenario that fits the facts and makes sense in terms of the motives of anyone and everyone involved. "Anticipation" requires us to imagine how the future might evolve from the present, based only on our previous experience of how the world works and whatever changes to those workings we can conceive of. And the implication of "most favorable" is that we can imagine multiple scenarios and multiple outcomes—multiple "possible worlds"—and then assess them to choose the one that best suits our proclivities and preferences.

Each of these faculties requires extraordinary cognitive capabilities: but they come so naturally to us that we're often barely conscious we're using them. In fact, we find it pretty difficult *not* to use them. If you ask people whether a glass of water is half full or half empty, they tend not to give you a straight answer. In and of itself, it's a boring, even meaningless question. Young children might start measuring the water level with painstaking care, but older kids and adults realise they're being tested and try to reason their way to an answer: if the water has just been poured, it's half full; but if it *was* full and you've drunk from it, it's half empty. Thinkfluencers on LinkedIn reframe the problem: the *real* point is that you can *decide* whether to refill or drain the glass. Jokers wisecrack about it: "An optimist says the glass is half full, a pessimist says it's half empty. And while they're arguing, a realist drinks it."

Psychologists certainly don't assess human optimism by presenting people with half a glass of water. So how *do* they do it? Well, for a long time, they simply didn't bother.

Sigmund Freud set the tone, suggesting that only the weak-minded needed to assuage their anxieties with irrational beliefs. Believing that things were bound to turn out for the best, perhaps through divine providence, might be a comfort to the superstitious and the unfortunate, but the well-educated and mentally robust would be content to take the world as it was. If people persisted in clinging to irrational beliefs when presented with contradictory evidence, they should be considered delusional, potentially in need of psychiatric treatment.

This attitude permeated psychologists' thinking for the first half of the twentieth century, but by the 1960s consumer research had revealed that *most* people expected their lives to go far more smoothly than they had any right to. These beliefs turned out to be held by a great many people leading perfectly unruffled lives. They claimed there was little chance they would be involved in a car crash or get cancer; they predicted that the winds of change in politics, economics and society would blow in the directions they favoured; they even expressed wholly unwarranted confidence about such entirely random feats as drawing a particular card from a shuffled pack.

The ubiquity of these beliefs—soothingly rebranded as "positive illusions"—was something of a puzzle. Could it really be the case that pretty much *everyone* was mildly delusional? Much chin-stroking ensued, but it wasn't until 1980 that someone undertook the first real study of this phenomenon.

Neil Weinstein, then a young Rutgers University psychologist, had been asking people to rate their chances of experiencing various unpleasantnesses like divorce, getting fired or being mugged. When entering their responses on computer punch cards, the data-crunching technology of the time, he noticed something odd: "All the responses were on the below-average side of the scale."

Intrigued, Weinstein asked a total of 130 students to rate their chances of experiencing 18 positive and 24 negative life events: college graduation, career milestones, marriage, divorce, illness. An important feature of his experiment was that he asked participants to compare themselves to others, not to the statistical likelihood of an event. As the consumer research had shown, people consistently overrate their chances of good fortune and underrate their chances of coming to grief. But then, not many people have a good understanding of their odds of getting cancer or being involved in a car crash. Making accurate estimates for those events requires a lot of data and actuarial expertise.

Asking people where they stand in comparison to others, however, doesn't require any objective assessment of the likelihood. So Weinstein's test measured whether his student subjects really were optimistic about their life prospects, not how bad they were at guessing statistics. And indeed they were: they believed they had an above-average chance of experiencing the positive life events and a below-average chance of experiencing the negative ones. There was considerable variation in how they rated different events. For example, there were about six times as many optimistic responses as pessimistic ones when it came to "owning your own home," but optimism only just nudged ahead when it came to "graduating in top third of class." In general, people seemed more optimistic when they viewed an outcome as more desirable, mundane or controllable.

And they *stayed* optimistic, even if Weinstein attempted to ground them. He asked one group to explain why they had rated their chances in the way they had, then showed them a compilation of the explanations given by other participants before getting them to take the survey again. The idea was to show them that they might not be exceptional in the ways they had assumed. Sure enough, the students did become more thoughtful and moderate in their responses—but *still* said they were less likely to experience negative events than average.

In the decades since Weinstein's pioneering experiments, literally hundreds of further studies—using a variety of methodologies and definitions of "optimism"—have confirmed that people are generally inclined to overestimate their chances of good fortune and underestimate their chances of misfortune. Unwanted pregnancy, the consequences of smoking, getting cancer, breaking up;

earthquakes, car crashes, even radon contamination in housing—people have been shown to be unrealistically optimistic about all of these and more. It's also true of more mundane events: we think holidays will be more fun than they turn out to be, and we expect more good times—hearty meals, fun nights out—than we actually enjoy.

The gulf between our expectations and reality is called the "optimism bias," and those hundreds of studies confirm that it's widespread, regardless of gender, race, nationality or age. "Even experts show startlingly optimistic biases; divorce lawyers underestimate the negative consequences of divorce, financial analysts expect improbably high profits, and medical doctors overestimate the effectiveness of their treatment," writes Tali Sharot, the neuroscientist who has done much to popularise the concept.

By Sharot's estimation, roughly 80 per cent of people exhibit the optimism bias to some degree or another, while about 10 per cent are realists, whose estimates are generally on the mark, and 10 per cent are pessimists, who tend to expect the worst. That last group is small but significant: it includes people with depression, for which pessimism is one of the diagnostic criteria. Understanding that association is one of the key motivations for psychological research in this area. But the majority is also worth investigating. Why on earth are most of us optimistic?

Unrealistic Expectations

For millennia, great thinkers have contended that the best way to proceed through life is to take as realistic a view of its opportunities and challenges as possible—where "realism" means taking a view of the world that is firmly rooted in evidence, logic and rationality. That surely gives you the best chance of anticipating what the future holds and planning accordingly. If you see storm clouds gathering, you wouldn't insist on dressing for glorious sunshine. It's a tenet of mainstream economics, for example, that accurate evaluation of perfect information is the ideal way to maximise reward and minimise risk. The assumption that realism is best is built into many of the ways we design and run our societies.

And yet in recent years it's become increasingly apparent that we don't actually adhere to this principle, in economics or anywhere else. The way our minds work, and the decisions and actions that result, are frequently out of line with what arithmetic, evidence or logic would dictate. Dozens of "cognitive biases" have been identified over the past half century, instigated by the pioneering work of Amos Tversky and Daniel Kahneman and popularised in the latter's bestselling book *Thinking, Fast and Slow*. Some have become well known, such as "confirmation bias"—the tendency to seek out, save and remember information that backs up our own prejudices while ignoring or forgetting contradictory evidence. We'll see later that this is similar to how we stay optimistic despite life's setbacks and failures: we cling to bits of information that support positive expectations of our futures and downplay those that don't.

We also use mental shortcuts—heuristics—to avoid having to solve complicated problems from scratch every time we encounter them. The "representativeness heuristic," for example, encourages us to think in terms of stereotypes: if I tell you that Doreen is quiet, shy, orderly and likes cardigans, you're more likely to guess that she works as a librarian than as a hockey player. But as you can probably imagine, given this example, heuristics can also lead us astray, with potentially regrettable consequences.

Another example is the "availability heuristic," the tendency to judge the likelihood of an event according to how easily we can remember a similar event. Again, this can be useful for making decisions quickly, but also misleading in the modern world. Plane crashes are exceedingly rare, but spectacular; car crashes are common, but don't usually make the news. The result: people are so afraid of flying that they opt to drive instead, although that's actually a much more dangerous option. Of course, we know flying is the safest form of transport because every aspect of aviation is designed to be as rational and systematic as possible, from how pilots are trained to the way aircraft are maintained. We've tried, whether explicitly or implicitly, to eliminate cognitive bias, and with it risk. We don't just wing it, if you'll pardon the phrase.

Should we try to extinguish the optimism bias, too? There's a tendency, particularly among some who fetishise rationality, to view cognitive biases as errors to be stamped out. That's particularly evident, and justifiable, in safety-first fields like aviation; but it's also the case in fields like economics, where the human tendency to act on emotion and instinct—"animal spirits"—has been treated as an irritating deviation from how conventional theory says people *should* act. But such abstracted perspectives obscure the truths of dealing with other people over the course of a lifetime, amid the messiness of the real world, where information is rarely perfect and people even less so. What looks foolish on paper, or in a laboratory, may make perfect sense in practice, or in a village.

We've already seen how optimism might work *for* us in the real world: by prompting us to action when the outcome is uncertain. In some situations, error management theory suggests that this can actually lead to better results than going by our most carefully reasoned estimates. But that works out in the average, over time: our positive illusions certainly can lead us astray on any given occasion. One notorious example is the "planning fallacy"—better known to my friends as the "Why is Sumit always late?" problem. Optimists like me apparently find it difficult to imagine that anything will impede our progress from A to B—whether negatives like a traffic problem, or positives like an unexpected chat with a neighbor—and so are perpetually running behind. That's true even when we *know* we're always late: we can acknowledge that we've been too optimistic in the past but insist that we're being realistic this time. Being late (usually) costs us little, however much it irritates our friends, so there's not much incentive to improve our error management. And so we end up being late yet again.*

With Tesla and SpaceX, Elon Musk revolutionised two industries against most people's expectations; secured a pay deal which made him the richest man in the world; and attracted legions of fanboys (and let's face it: they *are* boys) despite—or perhaps because of—what might generously be described as an idiosyncratic approach to running a business. "Who would try to do [self-driving cars] *and* rockets if they weren't pathologically optimistic?" he told Tesla investors in 2023. He's achieved extraordinary success, by many measures, against expectations that many would consider wildly unrealistic. His empire strongly resembles a personality cult, and one that's built around optimism.

But Musk also exhibits the downsides of extreme optimism, notably the planning fallacy. He's notoriously insisted every year—for more than a decade—that Teslas are *just* about to become fully self-driving. In 2018 he claimed that SpaceX would put humans on the Moon *and* a lander on Mars the following year—before the company's Starship had even flown. "I do have an issue with time," he told Tesla's shareholder meeting that year. "This is something I'm trying to get better at." Years later, Tesla's cars still don't drive themselves and SpaceX still hasn't put humans anywhere beyond low Earth orbit, while Musk has moved on to ruining Twitter while promising dancing robots and telepathic brain implants. Will he put a million people on the Red Planet by the 2060s, as he's promised? You wouldn't bet on it. But maybe you wouldn't rush to bet against it either.

* Once I became aware of this, I made heroic attempts to overcome it. I can't say I always succeed, but these days I more often get there in the nick of time than catastrophically late.

You probably should bet against most such "megaprojects," though. Grand construction schemes, ranging from the Sydney Opera House to London's Crossrail subway system, are notoriously prone to running way behind schedule and far over budget. The same is largely true of other kinds of megaprojects—corporate IT overhauls, for example, or engineering programmes like the development of new aircraft. In fact, that's the rule rather than the exception: 9 out of 10 megaprojects with a budget of more than $1 billion overrun or overspend (or both), sometimes catastrophically. Optimism bias means engineers imagine a project sailing along smoothly—and budget accordingly—rather than all the possible ways in which it might blunder into troubled waters. (And of course they are under pressure to come up with the lowest defensible bid and the fastest conceivable timeline, too.)

We make models, predictions and forecasts precisely so that we can avoid these kinds of fiascos, and evaluate our future prospects more objectively than our own intuition allows. Megaprojects are still a work in progress (to say nothing of Musk's leadership style), but we've done better in other areas. Computer simulations tell us whether it's likely to rain tomorrow, and how much the global temperature is likely to rise over the next century. We can estimate our general odds of getting cancer from population studies, or our individual odds from genetic testing. An actuary can tell us how likely we are to be involved in a car (or plane) crash. We may be inclined to ignore or flat-out deny these estimates, but in the end they exist to temper our expectations.

There are still plenty of things we can't predict, however. We might know our odds of cancer, but still can't say definitively whether we will ever be diagnosed with it, still less exactly when that might be. We have even less ability to predict a rare event like a car (or plane) crash, or for that matter any number of mundane experiences: meeting a partner, conceiving a child, getting a new job. In those cases, can it pay to look on the bright side, rather than taking a balanced view? It turns out that it can.

New Orientations

Weinstein's 1980 study opened up a huge new field of psychological research. But it was just a start. Asking test subjects about specific life events gave them a lot of scope to rationalise their answers—in much the same way that they might have done if presented with the proverbial half-glass of water. That could lead to confusing, potentially misleading results. A clean-living twenty-something might quite realistically say they were less likely than average to develop cancer in the next year, but a sixty-something chain-smoker might *un*realistically claim the same, on the spurious basis that their tobacco-addicted grandparent had died peacefully in their sleep at the age of ninety-eight. In that case, the sixty-something would actually be the more optimistic of the two, but Weinstein's approach couldn't pick that up.

This has significant implications. Later research showed that people who underestimate their risk of a particular kind of illness may be less likely to take precautions or heed advice against it. Unrealistically optimistic smokers, for example, underestimate their chances of getting lung cancer, and overestimate their ability to quit: "I can give up whenever I like" pretty much sums it up. The same sort of attitudes have been found in people at risk of heart conditions, of sexually transmitted diseases, alcohol misuse and gambling problems. Predictably, their optimism generally proves to be misguided; it may even foster behaviour that makes it so. It also raises tricky questions about their medical treatment: What does such an optimist really take on board when a doctor tells them there's only a one in ten chance that a procedure will work? Can they really be said to be giving their informed consent?

In 1985, Michael Scheier, at Carnegie Mellon University, and Charles Carver, of the University of Miami, decided to take a different tack. Like Weinstein, they had been investigating how people's expectations helped them achieve their objectives or overcome obstacles, initially in highly specific ways. For example, they asked people who were afraid of snakes to approach a caged boa constrictor while looking at themselves in a mirror, intended to help them literally reflect on their mental state. Sure enough, those who expected they'd be able to overcome their fear got closer to the snake than those who didn't. But how could such a contrived experiment be related to real life?

Scheier and Carver decided to investigate whether people might have a *general* expectation that things would turn out well in life, rather than a specific task or event. They came up with a dozen statements which they thought were representative of general attitudes, such as "In uncertain times, I usually expect the best," and "If something can go wrong for me, it will." People were asked how strongly they agreed or disagreed with these statements; their scores were then used to place them on a scale from low to high optimism. The test is pretty simple—in fact, you can try it for yourself in just a couple of minutes.

Here are the instructions provided for the most widely used version of the test:*

> Answer the following questions using the following scale: 0 = strongly disagree, 1 = disagree, 2 = neutral, 3 = agree, and 4 = strongly agree. Be as honest as you can throughout, and try not to let your response to one question influence your response to other questions. There are no right or wrong answers.

* Scheier, M.F., Carver, C.S., Bridges, M.W. (1994), "Distinguishing optimism from neuroticism (and trait anxiety, self-mastery, and self-esteem): a reevaluation of the Life Orientation Test," *Journal of Personality and Social Psychology*, Dec., 67(6), 1063–78: https://doi.org/10.1037/0022-3514.67.6.1063

	STATEMENT	SCORE
1	In uncertain times, I usually expect the best.	
2	It's easy for me to relax.	
3	If something can go wrong for me, it will.	
4	I'm always optimistic about my future.	
5	I enjoy my friends a lot.	
6	It's important for me to keep busy.	
7	I hardly ever expect things to go my way.	
8	I don't get upset too easily.	
9	I rarely count on good things happening to me.	
10	Overall, I expect more good things to happen to me than bad.	

You can actually ignore the scores given to statements 2, 5, 6 and 8—these are only there to obscure the purpose of the test. Statements 1, 4 and 10 are the ones that measure optimism; simply add up the scores given for those. Statements 3, 7 and 9 are negative, so *reverse* the scores given to these (that is, switch 0 to 4, 1 to 3, etc.), and add them up. Finally, add the two subtotals together to get your optimism score. A score of less than 13 indicates low optimism; 14–18, moderate optimism; 19–24, high optimism.

I had read quite a few studies that referenced Scheier and Carver's Life Orientation Test (LOT) before I looked it up. When I eventually did, I confess to being a little underwhelmed. Tests that ask you to score yourself are always a little suspect, since different people may interpret the questions quite differently, particularly when asked if they're "always" likely to feel a certain way or "usually" inclined to do something. (Does "usually" mean "nine times out of ten," or does it mean "anything more than half the time"?) The questions also seemed rather transparent. I scored the maximum 24 points, but wondered whether I was being honest with myself, or—as the author of a book about optimism—whether I had put down what I *thought* I should score. Most people, of course, are not so strongly motivated, but I also wondered if people living in cultures where optimism is generally celebrated (like the US) might feel a need to up-vote themselves, while those in countries where it's widely seen as gauche (like France) might downplay it.

But psychology is not an exact science. If you want to find out how people feel, you basically have two options. The first is to see how they react to a given stimulus and try to infer their mental state or thought processes from those reactions. Those experiments tend to be complicated and expensive to set up, particularly if they use high-tech tools like brain scanners, and their results are often difficult to interpret. The second is to just *ask* people how they feel. From that point of view, the LOT's simplicity was a distinct advantage. It was something of a blunt instrument, but then the virtue of a blunt instrument is that almost anyone can wield it.

"I think one reason the work was picked up so much is that we provided a tool that enabled scientists to ask their own questions and do their own research in the area," said Scheier in a 2012 interview with Hans Villarica for *The Atlantic*. "I think it also helped that our scale was easy to use and score. It only has six items on it! The brevity enabled lots of people to include it in their work, even if that involved very large epidemiological studies where issues of respondent burden and time limitations are paramount. As a result, an enormous amount of research on optimism has been generated over the years."

The very first of that research was published in 1985, when Scheier and Carver asked 141 undergraduates at Carnegie Mellon to complete both the LOT and a checklist of the most common symptoms of physical ill health, such as dizziness, blurred vision, muscle soreness and fatigue. The students first completed the questionnaires four weeks before the end of the semester, and then completed them again just before their final exams—a period during which one might expect them to be busy and stressed, and thus more subject to malaise.

Just as Carver and Scheier had hypothesised, students who scored more highly on the LOT reported being less troubled by the symptoms of ill health. The reverse was true for those who had scored lower. This, they suggested, was not necessarily because the high scorers actually experienced better health, but because they were better at coping with any illness or injury they did suffer. This has gone on to be a recurring motif of optimism research: optimism won't necessarily make outcomes better, particularly when they're largely beyond your control, but it will make you better able to deal with them without becoming stressed.

"Optimists are not simply being Pollyannas; they're problem-solvers who try to improve the situation. And if it can't be altered, they're also more likely than pessimists to accept that reality and move on," Scheier explained in 2012. "Physically, they're more likely to engage in behaviours that help protect against disease and promote recovery from illness. They're less likely to smoke, drink, and have poor diets, and more likely to exercise, sleep well, and adhere to rehab programs. Pessimists, on the other hand, tend to deny, avoid, and distort the problems they confront, and dwell on their negative feelings. It's easy to see now why pessimists don't do so well compared to optimists."

Higher levels of optimism have been linked to better sleep quality, lower inflammation and healthier levels of cholesterol and antioxidants; optimists have fewer heart problems and have been found to cope better with stress, pain, cancer and infertility. We've already seen that there's a relationship between pessimism and depression; optimists have been reported as more resilient in the face of extreme events like natural disasters and terrorist attacks.

The list of positive claims goes on, so much so as to provoke a certain amount of scepticism. You might wonder whether the LOT is *so* easy to administer that it has been attached to any and all studies, some of which are bound to be of lower quality than others: maybe psychologists themselves have unrealistically positive expectations about optimism's benefits. But the cumulative effect of all those studies, including some very large and well-controlled ones, suggests there is a real association between optimism and well-being. Overall, the conclusion is that optimists tend to enjoy longer, happier and healthier lives.

It's also easy to get the impression that optimism *brings about* well-being, particularly when you go beyond carefully worded research literature and into self-help manuals and online explainers (and certainly when you reach the mystic fringe of manifesting, where simply wishing hard enough is supposedly enough to make good things happen). In fact, the relationship between optimism and well-being probably isn't one way, but more like a circle, or a cycle. Common sense says that if you're in good health, have a great job and a satisfying personal life, you'll probably find it easier to look on the bright side. Biology may play a part, too: there's some evidence that some of the same genes influence both your level of optimism and your predisposition to good mental health. But being optimistic means you gain benefits you weren't expecting, which improves your standard of living, which validates your previous optimism and makes it easier to be optimistic in the future.

The cycle can be broken. People with depression score low on the Life Orientation Test and other measures of optimism, reflecting the extreme difficulty they have in anticipating positive outcomes from anything they do. From a clinical perspective, that's the real benefit of optimism research: it can help us understand, and thus improve, well-being. A counsellor acquaintance told me that what she finds most frustrating in talking to people with severe depression is how clearly she can argue that their situation would be improved through small, simple actions—but they simply can't be persuaded. Changing their perspective even slightly could be the first step to recovery—if they only believed it.

Agreeable Dispositions

Most people *do* believe in the power of optimism. In 2008, the psychologists David Armor, Cade Massey and Aaron Sackett asked 383 people to consider four hypothetical challenges: making a financial investment, organising a party, accepting an award nomination and deciding on a course of treatment for a heart condition. In each case, the researchers wanted to test how participants' evaluations changed according to three variables: the protagonist's level of commitment, decision-making agency and control over the situation. So, for example, the party organiser might have volunteered, or had the job delegated to them; might be in charge of the invite list or not; and might or might not decide how to spend the budget.

In all but one of the vignettes, they prescribed a modest degree of optimism as the best attitude, with greater optimism deemed appropriate if the protagonist had greater control over the outcome. But participants felt that in practice the protagonists would only be about half as optimistic as they "should" be: while they endorsed the value of optimism, they apparently also believed that not many people would fully exploit it.

Armor, Massey and Sackett didn't define "optimism" in their study; they just asked whether it was a good approach to take. That left a lot of room for interpretation. Did their students take "optimism" to mean an above-average chance of success, a better performance than previously experienced, an intuitive feeling of positivity, a determination to make a success of the challenge at hand or something else? Follow-up studies suggest that people express more nuanced attitudes when the questions are worded differently. While "people generally prescribe being optimistic, feeling optimistic, or even thinking optimistically," according to one study, they did *not* say it was a good idea to overestimate the likelihood of success. You might interpret this as endorsement of optimism as a general attitude, but not as an unrealistically positive illusion about a specific outcome.

Optimism is not one thing. Our expectations of the future vary with context. So do the ways we evaluate our likelihood of success when asked. The positive illusions uncovered by market research constitute *absolute* optimism: the belief that you will beat the odds. Weinstein investigated *comparative* optimism: the presumption that you will fare better than your peers. Both of these kinds of optimism are about a particular situation, and they're considered "unrealistic" because they can't be generally well-founded. Some people might beat the odds, but by definition not everyone—or even most people—will.

Scheier and Carver's Life Orientation Test, on the other hand, assesses *dispositional* optimism: a general, all-purpose expectation that things will work out well for us that doesn't apply to any particular situation. It's just an attitude that can encourage us to take measured risks, work towards our goals and enjoy greater well-being as a result. We'll encounter more kinds of optimism as we go along—some of them well-defined, some less so.

The distinctions between kinds of optimism can be important. One famous example is best known as the Stockdale Paradox, after the US Navy vice admiral James Stockdale, who was held as a prisoner of war in Vietnam for seven years, during which he suffered torture on a daily basis. How did he endure it?

"I never doubted not only that I would get out, but also that I would prevail in the end and turn the experience into the defining event of my life, which, in retrospect, I would not trade," he told Jim Collins in the leadership manual *Good to Great*. This might sound like the dictionary definition of optimism: irrationally positive expectations at the time, and the most favourable possible construction in retrospect. But Stockdale didn't consider himself an optimist: to him, the optimists "were the ones who said, 'We're going to be out by Christmas.' And Christmas would come, and Christmas would go. Then they'd say, 'We're going to be out by Easter.' And Easter would come, and Easter would go. And then Thanksgiving, and then it would be Christmas again. And they died of a broken heart."

Stockdale, understandably, didn't make this distinction, but we could argue that he's actually describing two different kinds of optimism. The first is his own extraordinary dispositional optimism, an open-ended expectation of eventual triumph; the second, his fellow PoWs' unrealistic optimism, pegged to specific timelines over which they had no control and were thus destined to end in recurring disappointment. Dispositional optimism keeps you going, prompting you to look for solutions and explore alternatives; unrealistic optimism *can* help you endure, but when you come up against an immovable obstacle, the disillusionment can be crushing.

Such distinctions aside, there's still a general statement which we can make: optimism is not the preserve of some ebullient minority of humankind. Instead, it's closer to being our default state. "Optimistic errors seem to be an integral part of human nature, observed across gender, race, nationality and age," as the neuroscientist Tali Sharot wrote in 2011. That does *seem* to be the case from the research we have—but much of that research was conducted on tightly defined groups of people. Often this is for good reason: for example, researchers seeking to understand who is optimistic and why might compare younger and older people; or they might focus on people who have had specific life experiences.

But sometimes—often—the research pool is defined by convenience. A disproportionate amount of psychological research is conducted on students and nurses, including many of the studies cited in this chapter, for the simple reason that they are time-rich, cash-poor and easily recruited in the universities and hospitals where the studies are taking place. Research also tends to happen only in places that can afford it, limiting geographical diversity. Harvard anthropology professor Joseph Henrich refers to the participants whose psyches dominate such research as WEIRD: western, educated, industrialised, rich and democratic. This prompts the thought that perhaps it's only WEIRD societies that are, or can afford to be, optimistic. What about people elsewhere?

One of the most international analyses conducted to date was based on a single question put to just over 150,000 participants in the 2013 Gallup World Poll, hailing from 142 countries: "Please imagine a ladder with steps numbered from zero at the bottom to 10 at the top. The top of the ladder represents the best possible life for you and the bottom of the ladder represents the worst possible life for you. On which step of the ladder would you say you personally feel you stand at this time? On which step do you think you will stand about five years from now?" There were significant differences between different countries' levels of optimism. Ireland, Brazil and Denmark reported the highest levels, while Haiti, Egypt and Zimbabwe reported the lowest. More than eight out of ten respondents worldwide said their life would be better five years from now; only in Zimbabwe was the average expectation that it would be worse. The poll also assessed how respondents rated their mental and physical health, finding that optimism was strongly linked with "positive affect" around the world, but that its relationship with negative affect, life satisfaction and perceived health varied from country to country, for reasons that remain unclear.

"The present study provides compelling evidence that optimism is a universal phenomenon, that optimism is associated with improved perceptions of physical health worldwide, and that optimism is associated with improved subjective well-being worldwide," the authors concluded. "Our results therefore suggest that optimism is not merely a benefit of living in industrialized nations, but reflects a universal characteristic that is associated with and potentially may serve to promote improved psychological functioning worldwide."

Until somebody undertakes a more comprehensive study, that serves as the definitive statement on the matter. Optimism is universal. Optimism is an illusion. And optimism is beneficial.

Social Contagions

So far, we've examined how optimism works in terms of our individual psychologies, the personal expectations they generate, and the idiosyncratic benefits that might follow. But human beings are intensely social animals. We live in couples, families, tribes, villages, cities and nations. Our intuitive optimism extends to those close to us: we believe they will lead lives that are more charmed than can reasonably be expected, although not to the same extent that we overstate our own prospects.

We may extend it to those beyond our inner circle, too, as long as we identify with them: we're optimistic about the success of the sports teams and political parties we support. Our willingness to extend optimism to randoms depends on how warm and competent we find them: broadly speaking, the more we like and respect someone, the more likely we are to be optimistic on their behalf. So optimism pervades our social relationships as well as our individual decisions, but does it actually influence them?

The answer might lie with those who've made influencing into a career. "Tapping through Palak Joshi's Instagram Stories recently," began a 2018 piece by Taylor Lorenz in *The Atlantic*, "you might have come across a photo that looked like standard sponsored content: a shiny white box emblazoned with the red logo for the Chinese phone manufacturer OnePlus and the number six, shot from above on a concrete background. It featured the branded hashtag tied to the phone's launch, and tagged OnePlus's Instagram handle. And it looked similar to posts from the company itself announcing the launch of its new Android phone." But as it turned out, the post wasn't an ad: it just looked like one.

Why? Because the appearance of success—which for an aspirant influencer means getting paid for posing with products—can beget success. Conspicuously thank a restaurant for its hospitality after paying the bill; caption selfies as though your paid vacation is some sort of junket; lean on a fancy car that someone else has parked outside a luxe store. Eventually your follower count will reach the point where people will start sending you stuff for real. (And if you need some help you can boost that count, too, whether with humans or bots.) "Fake it till you make it" has become a viable influencer career path—an intrinsically optimistic one, because to embark on this path, you have to believe that it will work for you.

It's been suggested that optimism is a way of deluding *ourselves* about our abilities and prospects in order to better fool others. If you genuinely believe you're a great catch, you're likely to present yourself more convincingly to a potential partner; similarly, if you really believe you're going to win a fight with a competitor, you're more likely to give it your all. That makes it more likely that you will end up with a desirable match, or on top of a power struggle: and that makes it more likely that you will win out in the long run. Simple computer simulations demonstrate that this can be a winning strategy: over time, optimists outperform pessimists.

It is also plausible that optimism is socially contagious. We've all felt buoyed and energised when in the company of someone who asserts, even without any real evidence, that we can overcome an obstacle or rise to a challenge—and we've often found that, once fired up, we can. It's the stuff of team-building exercises and motivational speaking. People like to hang out with optimists, which makes it easier for them to gain positions of power and, well, influence. Maybe this is the primate equivalent of the foraging mouse: in human terms, it's worth expending energy on self-promotion if it recoups social capital.

Put all this together, and you might expect optimists to be socially and economically successful. There aren't the same deep troves of evidence for this as for health benefits, but it still stacks up to a decent case that optimists do better at accumulating "status resources such as seniority, leadership, money, and possessions[;] and social resources such as relationships with friends, family, and romantic partners," as the psychologist Suzanne Segerstrom drily notes. Less drily, optimists are more likely to persist with education or employment, and thus prosper over time. They also work harder at personal and business relationships, which are likely to prove more enduring and satisfying than those of non-optimists.

In the round, then, optimism (in the broadest sense) seems to be good for us, and we seem to recognise that it's good for us, too. We endorse it in our approach to life's challenges, and it seems to help us resolve them. That makes it all the more puzzling that there's a widespread belief—backed up by opinion polls, market research and economic surveys—that optimism is in short supply when it comes to the *big* challenges we face.

In large swathes of the world, people express far less optimism about the direction of their country, or the future of the world in general, than about their own lives. In response, calls for optimism turn up incessantly in hand-wringing op-eds, in LinkedIn think-pieces, at Davos and at dinner parties. We must be more optimistic, they urge, if we're to deal with our problems. Positive attitudes are important. Passivity and pessimism will get us nowhere. And yet the optimism gaps persist.

This isn't especially hard to explain. It dovetails with our somewhat defensible belief that the more agency we have over a situation, the more able we are to steer it towards a positive outcome. That's not so true of, say, climate change. In the face of a changing planet, we're all vulnerable, all culpable and—with exceptions—all powerless. And what we collectively feel can have very real consequences—including the self-fulling prophecy of the pessimism trap.

2

Prophets of Doom

Is the world sliding into a pessimism trap?

Perched high on the Rosneath peninsula, about an hour's drive west of Glasgow, is a hamlet of converted shipping containers scattered across a densely wooded valley. Here, at Cove Park, artists come to decompress, devise new work and discuss the issues of the day.

I'd made the journey there, one cold but brilliant morning in November 2020, to attend a cultural symposium on climate change. Its aim was to formulate new ways of talking about the climate crisis, an artsy reaction to the diplomatic talking shop at the COP26 climate summit taking place at the same time in Glasgow. I was there more out of duty than conviction. As a Cove Park trustee, I wondered if it was really a productive use of our resources, time and energy. Cultural types chatting about what climate change "meant": Was this fiddling while the world burned? The symposium agenda, written in fluent art-speak, did not make me any more confident.

In the event, I was pleasantly surprised. Some of the talks—from oceanographers and archaeologists, lawyers and technologists, as well as artists and writers—were practical, some were conceptual. A few struck me as ineffectual. But they all introduced new ways of looking at climate change, a subject which had come to induce a kind of mental catalepsy in me, my thoughts cycling endlessly through parts per million of carbon dioxide, degrees of temperature rise and dollars of energy price—the variables in play at COP26. I was reassured that while climate change remained a vast problem, there was an equally vast reservoir of ideas for how to think about it and act upon it.

The talk that stayed with me most, however, wasn't one that made me feel optimistic. It was from Aka Niviâna, a young Inuk poet and activist. Niviâna's distinctively Greenlandic way of speaking in slow, considered phrases came in contrast to the patter I had been hearing the rest of the day. Climate change wasn't a looming catastrophe in Greenland, she said: it was a real and present calamity. Temperatures had hit unprecedented highs in recent years. Glaciers were retreating. The environment was transforming the Inuit's homes, reshaping the landscape, before their very eyes.

Niviâna was speaking to us over video link from Greenland—an unpleasant reminder of the long, lonely months of Covid-imposed telepresence we in the UK had only recently left behind. I wondered, though, if someone on the frontline of global warming might actually prefer this mode of communication to the carbon emissions of flying. Perhaps our release from purgatorial online meetings was just a hiatus; further pandemics were almost certainly on their way. Perhaps a video link was more responsible than flying for anything other than "essential business" anyway. Perhaps she was speaking from our future.

There is no denialism in Greenland. What there is instead is despair. Greenland has for decades had the highest suicide rate in the world, rising from an annual rate of around 29 per 100,000 people in the early seventies to a peak of 121 per 100,000 in the late eighties, before falling back somewhat in the past decade. Even more alarmingly, this trend has been driven by the young: in contrast to other countries' experience, the suicide rate is far higher in Inuit below twenty-five than in older people. Everyone in that former group had lost someone to suicide, Niviâna said: friends, family members, schoolmates, colleagues.

Why mental health struggles are so prevalent among young Greenlanders is sadly familiar: a combination of individual circumstances (mental health problems, substance addiction, socioeconomic deprivation) and the societal factors that feed them (lack of access to educational and employment opportunities). These in turn have been linked to modernisation policies that disrupted communities and families during the same period, including rapid resettlement from traditional villages to new towns, and from traditional semi-nomadic lifestyles to consumerist, industrialised modernity.

Today there's another contributing factor: climate change. The Inuit, especially away from the bigger towns, still place great value on hunting and fishing—in particular, on being able to travel across sea ice, sometimes still on sledges drawn by dogs. Travelling over the frozen sea is also critical to visiting friends and family. Now they fear that's becoming untenable. "We no longer understand the land here," Claus Rassmussen, a sledge-dog hunter, told the journalist Dan McDougall in 2019. "We don't trust this rock. We need ice. Now you can have rain in the middle of the winter, and it can even start snowing in August. The frozen sea is what we need. We take our dog sleds on to the sea ice and we cut holes and peer into the ocean and take the same halibut our grandfathers have lived on. This is the balance of the seasons. This is the past." But it will not be the future.

There was a word for this estrangement circulating at that gathering in Cove Park: "solastalgia," coined by the philosopher Glenn Albrecht. We experience homesickness when we're away from home; solastalgia is the distress we feel when we're *at* home, but find it distorted and unfamiliar. The weather turns in ways we don't recognise, the landscape becomes unfamiliar, our homes inhospitable, our communities' behaviours warped and changed.

Many Inuit could be said to be doubly solastalgic: removed from their traditional homes and lifestyles then, divorced from their environment now.

The Inuit wouldn't be the first people forced into a new way of life by climate change. On the contrary, it's been happening ever since humans first walked on Earth, from the Akkadians in Mesopotamia to the Anasazi in Arizona. In fact, it's probably even happened before in Greenland. Today's Inuit are not its first inhabitants; earlier cultures may have become extinct because of climate fluctuations that eliminated their food supplies. This time, of course, those fluctuations are the result of human activity, overwhelmingly taking place far from the Arctic. But you could delude yourself into thinking this an unfortunate problem, for an unfortunate minority, far from home. It certainly felt far from *my* home.

Until July the following year, that is, when a heatwave in the UK led to the temperature peaking at more than 40°C (104°F), a temperature hitherto not so much unfamiliar as unimaginable. The plants scorched and died; my neighbourhood's already modest population of animals and birds hid themselves away. The humans did, too. More than 20,000 people had died of heat stress across mainland Europe the previous year, but the full force of that heatwave hadn't reached the UK. Many of those close to me had never experienced heat anywhere near this extreme.

I covered the expansive, curtainless windows of my kitchen—a sweltering greenhouse of a room even during a "normal" British summer—with cardboard. Once the card was up, the room was cool and dark. Harsh sunlight poked through the cracks; when the tape adhesive melted and pieces of card fell down, I winced and hurried to stick them back up. It felt like being in hiding from some post-apocalyptic horde. The climate war had come home.

Overdramatic? As it turned out, yes. But the very prospect of dangerously high temperatures was alien, in a country more normally known for its Goldilocks climate: never too hot, never too cold. The British summer had become hostile, something to be shut out rather than luxuriated in. And it was only set to get worse. This, I realised, was the mild, politely British version of solastalgia. While the increasing prevalence of heatwaves and other natural disasters might be dangerous in the UK, in many parts of the world they will be lethal. In parts of India and America, two other places in which I feel at home, they already are. Some places—places currently full of human life—might well become uninhabitable.

And so I thought again about what Aka Niviâna had said, and wondered again if she had been calling us from our future.

The Malady of Infinite Aspiration

A couple of weeks after my visit to Cove, a press release arrived in my inbox bearing reassuring news: "There is no shortage of reasons for pessimism in the world today: Climate change, the pandemic, poverty and inequality, rising distrust, and growing nationalism. But here is a reason for optimism: Children and young people refuse to see the world through the bleak lens of adults," said UNICEF executive director Henrietta Fore. "Compared to older generations, the world's young people remain hopeful, much more globally minded, and determined to make the world a better place. Today's young people have concerns for the future but see themselves as part of the solution."

Fore was referring to a UNICEF survey, billed as the first of its kind, which had asked both young adults and over-forties from twenty-one countries how they saw the world. And indeed, the sunny top-line *was* that the average youngster was much more optimistic about the world than the average elder: in fact, people became 1 per cent less likely to believe the world is becoming a better place with every year of age. The most optimistic kids were in Indonesia and Cameroon; the generation gap yawned widest in

Japan, the US and Argentina, largely because of disproportionately gloomy oldsters. Just over half the kids worldwide thought they would eventually be better off than their parents; only about a third thought their economic prospects were worse.

But the sunshine dimmed upon closer inspection. In some countries, less than a third of the younger age group thought they would be better off—a ratio not much different to the over-forties. Which countries? The wealthy ones. Sixty-one per cent of youngsters in high-income countries said they would be worse off than their parents. The corresponding figure in lower-income countries was just 24 per cent. "Most young people in rich countries have resources and opportunities that are the envy of young people in developing countries," noted UNICEF. Yet they don't seem to feel good about it: in fact, they're also much more likely to report a worsening of mental health than their peers elsewhere.

Think about it for a moment, and that's not really surprising. Between the global financial crisis and runaway socioeconomic inequality, young people have had it tough for much of the twenty-first century. Beyond that, the inhabitants of rich countries make disproportionate use of global resources and create disproportionate quantities of emissions and pollutants. In many cases, standards of living have been upheld by the importation of cheap goods and labour from other countries, which is now slowing as those countries get wealthier. And underneath all that is the slow tick of environmental degradation and climate change. It might seem only realistic to assume that things must go downhill from here.

A downturn with no end in sight can crush the spirits. A 2023 survey by the Japanese cabinet office found that 1.46 million adults of working age were living as *hikikomori*—social recluses who engage in little social interaction and rarely, or never, leave home. This phenomenon first emerged during the nineties, Japan's "lost decade," when economic stagnation doomed the prospects of a generation, and it's a condition officially recognised by both medical researchers and the Japanese government. Various measures have been tried to encourage *hikikomori* to re-enter society, including everything from cash payments to buddy programmes and VR simulations of re-entry to the outside world. None has worked.

Japan isn't alone. In 2011, an article in South Korea's *Kyunghyang Shinmun* newspaper coined the term "the *sampo* generation," literally the "three giving-up" generation. The three things given up on were dating, marriage and children, widely understood to be because of the stagnation of the country's once vibrant economy, where the cost of living remains high but decent jobs have become scarce. The coinage was swiftly expanded to include greater degrees of apathy: the *opo* (five giving-up) generation additionally relinquished employment and home ownership; *chilpo* (seven) interpersonal relationships and hope of a better future; *gupo* (nine) health and appearance; finally, and most bleakly, the *sippo* (ten) gave up on life itself.

Other rich countries, too, are seeing their younger citizens opt out, albeit so far in less pronounced ways. "Goblin mode" became the Oxford University Press's 2022 word of the year, winning the public vote by a landslide: the lexicographers defined it as "a type of behaviour which is unapologetically self-indulgent, lazy, slovenly, or greedy, typically in a way that rejects social norms or expectations." To some extent, this is the stuff of ironic memes and commentariat tut-tutting; and to some extent it's the perennial eagerness of some older people to bemoan their juniors' supposedly antisocial ways and lack of work ethic. But given the Japanese experience, it could become a more serious concern in WEIRD societies whose demographies and economies stagnate.

Disinclination to participate in society, or adhere to its norms, has a name: "anomie," coined by the pioneering sociologist Émile Durkheim. Working in the late 1800s, Durkheim posited that modern societies gave rise to many different, often highly specialised, forms of work, with little of the shared understanding he assumed to exist among the smaller and more tightly knit communities of a "primitive" society. This meant that people would develop very different perspectives on the world, weakening their adherence to social norms and leading, in times of dramatic change, to a sense of profound dislocation: anomie.

Durkheim cited the dismay experienced by craftsmen made redundant by industrialisation as an exemplar of anomie. Suddenly divorced from their lifelong purpose, and with a lifetime's investment in developing specialisms abruptly rendered worthless, they were directionless and despairing. They suffered, Durkheim wrote, from "the malady of infinite aspiration," a condition arising when people "no longer know what is possible and what is not, what is just and what is unjust, which claims and expectations are legitimate and which are immoderate." That, perhaps, is akin to what the solastalgics and the *hikikomori* are feeling.

Perhaps it's what the rest of us are beginning to feel, too. Online messaging and delivery services made it possible for Japan's hermits to never leave their rooms; increasingly, the rest of us have had our in-person social interactions mediated or replaced by screentime. The unexpected correlate to convenience and connection has turned out to be loneliness, which has now reached epidemic proportions—and "epidemic" is indeed the word, because it has been flagged as a global public health priority by the World Health Organization. A mass of circumstantial evidence suggests that being digitally native can be damaging to well-being; repeated waves of technological disruption, most recently from artificial intelligence, mean that people no longer know what is possible and what is not.

The consequences of anomie can be devastating. In his 1897 book, *Suicide: A Study in Sociology*, Durkheim set out to describe how taking one's own life could be viewed as a response to societal pressures and expectations—a stark contrast to the then-prevalent attitude that it was the result of an individual's own failures. One way this manifested, he suggested, was as "anomic suicide," brought on by an overwhelming change that leads to social dislocation and a profound sense of personal despair. I'm reminded, once again, of the Inuit.

Confined to quarters, doomscrolling, hiding from the sun. One way to account for the emergence of such widespread anomie is that the pace of change today—as well as the *perceived* pace of change today—has been so great that people are struggling to keep up. Because this change disrupts their positive expectations of the future, they retreat from society, a withdrawal from which it's almost impossible to escape. A pessimism trap. They are like mice too timid to venture from their burrows to find food or mates. Their error management has failed.

Blah, Blah, Blah

Cove Park is set between two lochs. The viewing platform in front of its central hub affords a sweeping vista across Loch Long: evergreen slopes, blue-grey water and off-white cottages. Trek up the hill path behind it and eventually Gare Loch comes into view, a reflection of its western sibling except that it's marred by the thoughtless beige sprawl of His Majesty's Naval Base, Clyde. From this vantage point, you can just about see the bays where Britain's four nuclear-armed submarines come into dock after their silent three-month tours of duty. When I had passed it earlier that day, the bays had been empty. The subs were out somewhere under the water, waiting for the declaration of World War Three.

That was a declaration I had spent much of my youth waiting for. As a tweenager, I had been terrified that the world would end in nuclear fire before I ever reached adulthood. It was a standard premise of the books I read and the films I watched (and those I hadn't watched but only heard about in the schoolyard, which was much worse). The news carried footage of female protestors camped at Greenham Common nuclear base; older kids went on marches and drew the peace sign on their school bags in felt-tip pen. For all that, discussing our anxieties was still somewhat taboo. During one class we were asked (for some reason) to write down our biggest fears. I expected to be one of a few who put "nuclear war" above "not passing my exams." It turned out pretty much the entire class had.

And yet, as time went on, my fear waned. The Cold War had thawed, with the Soviet Union first liberalising and then disintegrating. Amazingly, thanks to disarmament treaties, the number of nuclear weapons dropped significantly during this period. Various apparent flashpoints had come and gone without the use of nukes ever seeming plausible. While Armageddon was still entirely possible, not many people seemed that worried about it.

But some still were. On my way back to Glasgow from Cove Park, I passed the Faslane Peace Camp, an assemblage of dilapidated caravans tucked into the side of the road near the naval base—all shonky rainbow paint schemes, hand-lettered signs and tattered banners. It seemed deserted, but I knew it still had residents: some people, it seemed, still cared about averting nuclear war. What kept them here, long after most of their fellows had given up the struggle? Were they still motivated by fear of the bomb? Did this hamlet of protestors really hope to deter the activity behind the electric fence and armed guards on the other side of the road? At least they were doing *something* for their cause, I supposed. That was more than I could say.

Disembarking from my train into Glasgow's George Square, I was plunged immediately into what seemed to be a much bigger, and far more populous, version of the Faslane Peace Camp: vivid colours, crudely written placards, elaborate costumes and the occasional musical troupe. Some of the thousands of protestors in George Square that afternoon were the usual suspects: fringe politicos scouting for recruits, conspiracists touting photocopied screeds, anarchists awaiting their big moment. But most of them were kids skipping school, inspired by the teenaged activist Greta Thunberg, who was due to address the rally at the end of the day.

Thunberg's decision to "strike" from going to school in protest at political inaction on climate had made her a luminary, an almost saintly figure to many environmentalists. When I had first heard about it, I was surprised it had provoked any reaction beyond a weary note from the school to her parents. But in fact it had captured the imagination of other terrified youngsters, ultimately mobilising millions in her cause. It was a good example, I thought, of optimism without sunshine, without forced grins, without "positivity." To me, her action represented the triumph of "unrealistic" expectations over apathy and indifference. I was looking forward to hearing what she said.

The on-stage programme was nominally reminiscent of what I'd heard at Cove Park. Speaker after speaker from various walks of life, and from all over the world, told the crowd how the changing climate was affecting their lives. But while the speakers at Cove had probed at the problem, trying to find new ways to articulate it, understand it and ultimately tackle it, the ones in George Square seemed only to want to restate it. I was taken aback by how, even as they called for action, they seemed to deny, over and over, that any action could be taken. The situation was irredeemable. The politicians were useless. No one was listening.

Greta, ultimately, was no different. "The people in power can continue to live in their bubble filled with their fantasies, like eternal growth on a finite planet and technological solutions that will suddenly appear seemingly out of nowhere and will erase all of these crises just like that," she told the crowd. What they should be doing, she said, was enacting "immediate drastic annual emission cuts unlike anything the world has ever seen." The talks down the road were a failure (their outcome was not, at that point, actually known) that she dismissed angrily as so much "blah, blah, blah."

I was also pretty cynical about the COP process, to be sure, but it wasn't clear to me how else those drastic emission cuts were going to be enacted. And I was frustrated by the environmental movement's frequent disdain for technology. While politicians had been mithering, scientists, engineers and industrialists had brought about a quiet revolution in solar- and wind-power generation. That wasn't by itself enough, but it was the only substantive action anyone had taken—and to my mind, the more the politicians vacillated, the *greater* the need for technological solutions became. It was, for me, a disappointment to hear someone who had built such a platform use it to shout empty slogans.

But I also sympathised. I had felt equally furious and impotent at the failings of authority when I was a teenager confronting my own existential terror, and incapable of seeing a way out of it. The previous year, a BBC survey had found that one in five British children had had a nightmare about climate change. Two out of five did not think adults were doing enough about it; two out of three felt leaders weren't listening to their concerns. If they were pessimistic, it was the fault of my generation. The youthful protestors were convinced that they would not live to adulthood, or that if they did, they would be forced to scratch out a miserable existence in a post-apocalyptic wasteland—just as I had been at their age by the threat of nuclear Armageddon.

Nuclear weapons represent a pessimism trap that has run its course. All those rallies, all that paranoia, all those terrifying TV movies: all of that teenage terror. There are still more than enough nuclear weapons to destroy the world many times over, and that remains a highly plausible outcome if one is ever fired. Numerous opportunities to disarm completely came and went during the last third of the twentieth century. And now the logic of mutually assured destruction has dissolved: nuclear weapons are no longer a deterrent to war between the great powers, but the ultimate trump card in geopolitical negotiations, allowing anyone who has one to act with impunity. We did not believe a world without nuclear weapons was possible: now we're stuck with them, a perpetual, gnawing, existential threat.

The symposium still fresh in my mind, I wondered if concern about the climate would similarly recede from public consciousness as we discussed it, and internalised and normalised it. David Wallace-Wells, author of *The Uninhabitable Earth*—a book which started its life as the most widely read article in *New York* magazine's history—wrote that "What looks like apocalypse in prospect often feels more like grim normality when it arrives in the present." Were we in danger of giving up before we had even tried? Were we about to get caught in the ultimate pessimism trap: the lone frog drowning in milk?

Is It Okay to Still Have Children?

Not many men choose to be vasectomised at twenty-five, and most of those who do have already had as many kids as they want. Les Knight is an exception, opting for the procedure when he reached his quarter century in 1973. *He* had the snip to save the planet.

An ardent environmentalist, Knight was a member of the Zero Population Growth organisation, co-founded in 1968 by the biologist Paul Ehrlich. That same year Ehrlich and his wife, Anne, had published *The Population Bomb*, a bestseller which argued that the booming global population would result in widespread famine if drastic measures were not taken to restrict births. Over the next few years, evidence mounted of the pressure humans were putting on the natural world. In 1972, the Club of Rome, a kind of proto–World Economic Forum of thinkers, tycoons and wonks, published *The Limits to Growth*, reporting alarming predictions made by a pioneering computer simulation of the interactions between population, food production, industrialisation, pollution and resource consumption. The bottom line: Without a sharp change of trajectory, civilisation would collapse within a century.

By the late 1980s, environmentalism was becoming a force to be reckoned with. Calls to save the rainforest appeared everywhere, from kids' cartoons (*Captain Planet*) to shopping malls (the Rainforest Café). Images of oily seabirds and bloodied seal pups shocked animal lovers. A giant hole appeared in the ozone above Antarctica. Well-intentioned, if not always well-informed, campaigns abounded: against animal testing, nuclear energy, genetic modification, pesticides and more.

But while others were awakening to the plight of the natural world, Les Knight had already concluded that even zero population growth wasn't enough to fend off catastrophe. Only if people allowed themselves to die out gracefully would the rest of the planet's inhabitants survive and thrive. So in 1991 he founded the Human Extinction Movement, to argue for a stop to all procreation; on second thought, he added "Voluntary" to the name and abbreviated it to VHEMT, pronounced "vehement."

To spread the word, he started a newsletter, initially mailed to a few hundred supporters. "*These EXIT Times* hopes to voice the concerns of all of us who have volunteered to live long and die out," began the first issue. "As Volunteers know, the hopeful alternative to the extinction of millions, probably billions, of species of plants and animals is the voluntary extinction of one species; Homo sapiens . . . us." A few years later he set up a website on which the extinction-curious can find answers to questions like "Don't you like babies?" (they do) and "Didn't Hitler have the same ideas?" (he didn't); handy tips and hints (if your partner wants a child, try a realistic baby doll), and a video called "Thank You for Not Breeding."

Human extinction might not strike you as a terribly positive ambition, but Knight has been adamant since the very beginning that he sees it as an optimistic project. Its materials are cheerful, and focus on the promise of a brighter post-human future rather than the despair expressed by some other groups. He also acknowledges that his project is unrealistic: "Do you really expect to be successful?" is one of the FAQs. Knight's answer: "It has been suggested that there are only two chances of everyone volunteering to be VHEMT: slim, and none." Every person who hears and heeds the message represents a success, he says.

Initially, not many did. What attention VHEMT attracted during its first few years was at best amused and at worst derisory—it had perhaps a few hundred sincere supporters. Otherwise it was regarded as a crank organisation akin to UFO cults and flat-Earth societies. That didn't bother Knight. "I don't mind being considered a kook; somebody's got to do it," he told *The Economist* in 1998. "This is the natural progression of ideas. First we have to be ridiculed." In time, presumably, the idea would begin to be taken seriously.

Indeed. In 2019, thirty-one years after the Ehrlichs published *The Population Bomb*, the global population had reached 7.7 billion. Alexandria Ocasio-Cortez, born in 1989, had just become the youngest woman ever to take a seat in the US Congress, and she had a question for her 2.5 million Instagram followers.

"Is it okay to still have children?" she asked almost off-handedly during a livestream filmed from her kitchen.

It was a legitimate question, she said, and one that an increasing number of young people were asking themselves in the face of what she described as "scientific consensus that the lives of children are going to be very difficult." But the reason for her concern wasn't population growth. It was climate change.

For most of human existence, the question of whether it was okay to have children would have been virtually meaningless. Contraception was non-existent or unreliable, abstinence even more so. But the question of how to feed an extra mouth is eternally meaningful.

How many of the millions of people who become parents every year, no matter how carefully they think it through, truly know in advance how they're going to raise and care for their child? How can you guarantee your child's life will be a good one? Nothing about having a child is certain: not their health and happiness, not your ability to parent or provide, and certainly not the future they'll inhabit.

Having a child, with its life-changing consequences for all concerned, is thus one of the ultimate expressions of optimism—of "unrealistically" positive expectations of the future. Raising children takes a *lot* of energy, after all, for no immediately apparent return. Opinion surveys rarely ask "How optimistic are you?"—a nebulous question that invites subjective answers. Instead they ask "Will your children have better lives than you?" or "Will you have a better life than your parents had?" In this most biological of human affairs, we most closely resemble the animals we so often pretend not to be—it's hard to get animals to reproduce if they're stressed, as any zookeeper knows. When everything we hear about the world is alarming and everything we experience is unfamiliar, as we move through intolerable heatwaves and unseasonal downpours, it's no surprise that doubt sets in.

Some of the inhabitants of any given ecosystem are particularly sensitive to change. That makes them of particular interest to ecologists, not because they're necessarily any more important than the others,* but because they are indicator species whose fortunes reflect the health of the ecosystem as a whole. Corals, for example, are very sensitive to the temperature of seawater and become bleached when it's too high; frogs and toads absorb toxins through their skins, so are often the first to fall ill from pollution; some owls need old-growth forest to thrive, and vanish when those habitats are degraded.

* Some species *do* have a disproportionate impact on their host ecosystem: they're known as keystone species. A single pair of beavers can completely transform their environment, for example.

By way of analogy, Tim Gill, an advocate for children's play and mobility, describes children as an indicator species for society. "The visible presence of children and youth of different ages and backgrounds, with and without their parents, in numbers, is a sign of the health of human habitats. Just as the presence of salmon in a river is a sign of the health of that habitat," he wrote in a 2017 blog post. But "when parents today look out from their front doors, they see a world that is at best uncaring about their children, and at worst hostile to them."

And when they look into the future, they see the same. In 2020, just over 600 Americans of parenting age replied to a survey asking what they thought, and how they felt, about having children. The survey was unusual in that it attracted participants by asking prominent climate activists to publicise it, and gave respondents scope to write out their thoughts. Almost all of the respondents said they were either "concerned" or "very concerned" about how children would fare in a climate-changed world. Many said they simply wouldn't have children. "I don't want to birth children into a dying world. I dearly want to be a mother, but climate change is accelerating so quickly, and creating such horror already, that bringing a child into this mess is something I can't do," wrote one thirty-one-year-old respondent. "I feel like I can't in good conscience bring a child into this world and force them to try and survive what may be apocalyptic conditions," wrote an "undecided" twenty-seven-year-old project manager.

The participants in this survey could not by any stretch be described as representative, given how they were recruited. But the following year, a randomised survey of 10,000 young adults in ten countries, published in the august medical journal the *Lancet*, found that more than eight out of ten were at least moderately worried about climate change; nearly half said it was affecting their daily functioning, mainly because of their political representatives' failure to act. And two out of five said they were not sure it was a good idea to have kids. Other surveys have confirmed the strength of feeling.

The trepidation hasn't gone away. In an August 2023 piece for *The Cut*, the writer Amil Niazi reprinted some of the responses she got on social media when she asked Ocasio-Cortez's question. Among quotes from those yet to have children, but paralysed by anxiety, was one from Nadia Mike, an Inuk mother from Nunavut in northern Canada: "I have a 16-, 11-, and 5-year-old, and it's so sad to see the world deteriorating right before our eyes and to try to explain that it's fucked up, but also try to hold promise for them." Mike's brother had fled Yellowknife earlier that month, one of thousands forced from their homes by wildfires.

There's scope to wonder how far climate anxiety is serving as a lightning rod for a much broader spectrum of anxieties about having children, some old and some new. Finances are obviously a major factor, as are age-old worries about responsibility. As for new concerns, there's also the biodiversity crisis, habitat loss and pollution; socioeconomic inequality, corruption and polarisation; the hardy perennials of bigotry, sexism and intolerance; economic doldrums, bullshit jobs and automated redundancy; mental and physical health crises; autocracy, war, violence and death.

Clearly not everyone who chooses not to have children is gripped by existential dread: some just don't want to, and don't feel socially obliged to, as they once might have done. But it's also clear that many people just can't decide—or don't know *how* to decide—if they should become parents, given the vast uncertainties swirling around us today. It's as if they're looking endlessly at a glass of water, trying to decide if it's half full or half empty.

Easier to Imagine the End of the World

Some would argue that it's half full. Standards of life for most people on the planet have actually been getting *better*, by many important measures, from child mortality to household income. This, too, can stop people from having children—but because people become *more* optimistic, not less.

The "demographic transition model," a standard in population studies, begins with the observation that improvements in public health and sanitation lead to falling death rates, particularly among children, while new technologies make economies and societies more resilient. That makes people more confident about the future—which is to say, more optimistic. As they acclimatise to this more secure and affluent existence, they have fewer children, for reasons that include the availability of contraception; a move away from subsistence lifestyles that require large families to work the land; the education, emancipation and employment of women; and greater investment in education and careers.

And that's almost what happened: while the population has continued to grow, the *rate* of growth has decreased over time. People all over the world are having fewer children—not just in industrialised countries, but in Asia and the Global South, too. I say "almost" because the trend has gone further than expected. Under the original formulation of the demographic transition model, both birth and death rates drop until a new, but lower equilibrium is reached. But in fact birth rates have continued to decline rapidly. In 1960, the average South Korean woman had 6 children; today, that number has fallen to 0.78—well below the 2.1 "replacement rate" required to maintain a given population. That's acute, but by no means unique. According to one study, the populations of 23 nations—including Japan, South Korea, Italy and Spain—will plummet to half their current levels by the end of the century. In 1950, the global birth rate was 4.7 children per mother; today it's 2.4, and falling.

That, along with increasing longevity, means that most of the world's countries will become less populous in the second half of this century, and those populations will be "greying," with an increasing proportion of the elderly and retired. That brings inevitable and significant social, economic, political and cultural stresses.

So far, this has been partially addressed by supplementing the native population with youthful immigrants; but that may not remain viable, particularly since the countries that have supplied much of this labour—notably in Southeast Asia and Africa—are themselves passing rapidly through the stages of demographic transition. The economic load on younger people will make the already daunting cost of raising a family even more of a deterrent.

Some people aren't having children because they feel optimistic about the future; some people aren't having children because they feel pessimistic. But in sum, the world's next big problem will be that it has too *few* people to support its societies, not too many.

Various remedies have been proposed for this situation. Conservatives appeal for a return to supposedly "traditional" family values, painting child-rearing as patriotic, enlarging access to fertility treatment and restricting reproductive rights. These measures conflict with individual liberty, personal choice and changed social norms, and they don't work anyway. Liberals have suggested making working parents' lives more comfortable with tax breaks, childcare, free transport and other incentives. South Korea, which has declared its demographic decline a national emergency, spent more than $270 billion on incentives for people to become parents between 2016 and 2023. That hasn't worked either.

Of course, one could argue the state—a democratic one, anyway—has no more business trying to make people have children than it does stopping them. A government is meant to provide for the wants and needs of its population, not the other way round—even if that means the country hollows out.

If you take that view, you have to look to social and technological innovation for solutions. Perhaps older, healthier people can be persuaded to stay in work, with education and employment retooled to ensure their skills remain useful, as my London Business School professors Lynda Gratton and Andrew Scott described in their book, *The 100-Year Life*. Or perhaps artificial intelligence and robotics will hugely increase productivity and do the jobs for which no human is available, as some Asian nations expect. Or it might be a problem that solves itself. Perhaps people will just start having more children again if and when the world begins to seem a calmer, less crowded place.

The Japanese village of Ichinono, north of Osaka, had become so moribund that it had put up life-sized dolls of families in order to feel more populous. But that changed with the arrival of Kuranosuke, the first child born in the village for more than twenty years. "On earth for a little over a year and cherished by a cooing, tribute-bearing platoon of surrogate grandparents from around the village, the boy has already had poems written about him," reported Leo Lewis in a piece for the *Financial Times* that's both touching and eerie. "When he appears to his elderly local fans, he is the centre of attention and the target of collective parenting muscles, unflexed in some cases for over half a century." It takes a village, as they say; Kuranosuke has one.

With so many disparate possibilities in the air, it's no wonder that it's *hard* to decide if it's all right to still have children. Perhaps the best answer I've heard to Ocasio-Cortez's question came from the writer Meehan Crist, in an eloquent and impassioned talk at the British Museum on Valentine's Day, 2019. She argued that the responsibility for the climate crisis resided with the fossil-fuel industry, and specifically with BP—major sponsors of the museum hosting her talk, and the inventors of the idea of the "carbon footprint," a successful bid to shift responsibility for climate change to individuals, and a greenwashed twist on the privatisation of reward and the socialisation of risk. The piling of guilt onto individuals was, she argued, a deliberate deflection of blame by the forces of global fossil-fuel capitalism.*

* Decades later, they were still at it: in March 2024, the boss of Exxon—a company whose history of cover-ups and deliberate denial means it has more to answer for on climate change than any other on Earth—declared that price-sensitive consumers were to blame for its reluctance to cultivate greener energy sources.

We bought it. "It is easier to imagine the end of the world than to imagine the end of capitalism," wrote the critical theorist Fredric Jameson. We have to try, though: if not to imagine the end of capitalism, then at least to believe in a better world—because that's how we'll find our way to it. One individual can't possibly predict, in any detail, how their choices are going to affect the world. That's when we need optimism: to help us believe that our efforts are making a difference. For an increasing number of people, however, it seems that it's indeed easier to imagine the end of the world.

As You Face Annihilation

If Greta Thunberg was the big draw on that Friday afternoon in Glasgow, Extinction Rebellion (XR) was the big driver. Their X-in-a-circle symbol was everywhere: on flags, on banners, on T-shirts and stickers—a mash-up of the trident peace symbol and the anarchists' A-in-a-circle that neatly summed up where XR was coming from.

The movement was founded by veteran campaigners whose "Declaration of Rebellion" claimed the government had failed to act appropriately on the environment, and that citizens therefore had not just a right but a duty to rise up. It struck a chord. Within a year, XR's anarchic, carnivalesque protests had captured public attention like no other. When they occupied Central London's bridges and arterial roads in 2018 and 2019, for example, they pissed off a good many commuters, but also attracted a striking amount of support from the young and the frustrated. And from me: I believed *something* was needed to break the stalemate between ineffectual activist hand-wringing and cynical political windbaggery. But as time went on, I began to wonder about the apparent nihilism, not to say outright pessimism, of some in the movement.

XR co-founder Roger Hallam had previously been an organic farmer, and blamed extreme weather events for the demise of his small-holding; in response, he returned to academia to study radical protest and civil disobedience. His manner, in person, is mild, self-deprecating and straight-talking, with flashes of dorky humour. That makes the bluntness of his "Advice to Young People, as you Face Annihilation"—originally a kind of how-to guide for aspiring protestors—all the more startling. Issued in June 2021 as a Google document and a YouTube talk, it was written while Hallam was serving time in prison for planning to use drones to disrupt flights at Heathrow Airport.

In its most notorious passage, on the downfall of civilisation, he warns that climate change will cause societal breakdown—and when it comes, it will be swift and shocking. One day the bread will just disappear from the supermarkets, he suggests; the next day, someone will pay £50 for a loaf of bread in the car park; the next, there will be dead bodies as those starving begin killing for food. And then, he suggests, "A gang of boys will break into your house demanding food . . . They will see your mother, your sister, your girlfriend and they will gang rape her on the kitchen table. And after that, they will take a hot stick and poke out your eyes and blind you. That's the reality of the annihilation project you face."

Hallam and other prophets of doom claim to be speaking a truth that no one else will—not politicians, not the media, not even other environmentalists or climate scientists, for reasons of dishonesty, cowardice or venality. His stated objective is to shock young people into action. I think he and his ilk are passing off the most extreme possible outcomes as inevitable, making no acknowledgment of the progress that's been made or, more importantly, the progress that might yet be made.

So far, formal research is ambiguous on whether fear or hope is better at motivating action on climate. Fear and anger seem to be good ways of getting people worked up about a problem, but not particularly useful for getting them to actually *do* anything. Hope, on the other hand, seems to encourage people to change their lifestyles in more enduring ways. And as we've seen, optimism tips the scales towards action—which can help us find new and unsuspected solutions to our problems. Prophecies of doom, on the other hand, can have a sere and steely-eyed "realism" to them—a kind of dark glamour, almost—but are ultimately self-defeating. Like Les Knight's Voluntary Human Extinction Movement, they represent a pessimism trap.

"Doomerism" has spread far and wide, abetted by high-profile media like Wallace-Wells's *The Uninhabitable Earth*, and also, of course, by the ceaseless unspooling of genuinely unsettling news about the environment, freak weather events, continuing political inaction and continued corporate intransigence. Talk of societal collapse thrives online, in despairing Reddit forums and gloomy creative collectives and as the memetic punchline to a million social-media exchanges about the future. It has become a subculture: we doomscroll through doomerist memes and doomerist blogs; there are doomerist influencers; doomerist TV shows; doomerist festivals and doomerwave music.

But doom isn't what the professionals talk about. When *Scientific American* asked six climate scientists what they would tell their kids about the future they faced, none of them offered anything like Hallam's cautionary tale. Of course, that's not surprising: What parent would tell their child that *that* is what awaits them? But they didn't demand radical action either. What they offered instead was reassurance: The future will be different. There will be challenges to overcome, but we can all play our part in overcoming them.

That remains the message of many in the environmentalist movement, including Christiana Figueres, the diplomat who led the breakthrough 2015 Paris Agreement on climate and espouses "stubborn optimism." "We don't have the right to give up or let up," she said in 2020. "Optimism means envisioning our desired future and then actively pulling it closer. Optimism opens the field of possibility, it drives your desire to contribute, to make a difference, it makes you jump out of bed in the morning because you feel challenged and hopeful at the same time."

Even a whistle-blower like David Wallace-Wells acknowledged a few years later that his outlook had become more positive. Asked in 2021 about having a child while writing *The Uninhabitable Earth*, he replied: "I guess the first thing I would say is, I don't think that global warming is likely to proceed at a level that makes life miserable for my children, or indeed, most people on the planet. I think that there is going to be an enormous challenge of responding to this crisis. It will introduce a lot of pain and suffering, but that alongside that . . . we will come to accept conditions that are much more brutal, climate-wise, than the ones we live in today, without feeling like the world has ended."

By 2023, Extinction Rebellion, suffering growing pains and facing increasing scepticism over its more disruptive activities, had parted company with Roger Hallam and was trying to strike a more constructive tone, appealing for unity with the more moderate groups it once disdained. I wondered how XR's stalwarts felt about this, and what they really believed. For all the bleak messaging I had seen from them, I presumed they must also believe in a way forward: Why else put your livelihood, and your liberty, on the line to protest?

Clare Farrell, one of XR's co-founders, had recently been in court on a charge of criminal damage after smashing the windows of HSBC bank. It looked like an open-and-shut case, but she was acquitted by a jury after explaining why she felt her action to be justified.* "It's painful for me to be part of a society so immoral, so off track," she told them. "It is set to destroy the next generation,

* This form of defense was subsequently banned. Hallam defied that ban in a later case over a protest that obstructed a motorway and was sentenced to five years' imprisonment. The judge said he had "crossed the line from concerned campaigner to fanatic," but both ban and sentence were widely criticized as excessive.

and billions of lives are likely to be lost, and my heart asks me to do the work which has the best chance of effecting a change of course." I wanted to know what that change of course might be. So I took tea with her.

Farrell, a sustainable fashion designer by profession, arrived dressed in head-to-toe neon; as director of the XR artists' corps, she had been responsible for much of their distinctive look. When I explained my optimism project, she laughed. "I'm the nemesis of all that," she said. "XR was designed to destroy false hope." True to her word, throughout our conversation she remained adamant that we were headed inexorably towards catastrophe: "We've gone from 'Something's going to break' to 'It's broken, it's done, we've fucked it.'" A lucid defender of XR's original hardline approach, she was bluntly scornful of anyone who didn't share its doomerism, whether other activists, environmentalists or technologists—never mind politicians, business people or (gulp) the media. Nor did she accept that this led to nihilism; on the contrary, she said, "I've seen so many people become so much more pro-social once they've faced the truth and got over their initial grief."

I wondered how Farrell had come to this position. Upheaval during her childhood had left her convinced that "life can fuck you any time," she told me. Did she think her experiences had made her more of a realist than other people? "Absolutely," she answered, without a moment's hesitation. For a moment, I wondered if she really was the realist and I was just clinging to my illusions. I respected her arguments for direct action, I empathised with her feelings of melancholy. I was awed by her strength of purpose and the courage of her convictions. But I also found her dismissals of different opinions peremptory and her arguments contradictory. When I eventually asked her what world she'd like to bring about, her answer was as nebulous as her doomerism had been decisive: one in which people were, in some way, more cooperative and thoughtful in their dealings with each other.

I was trying hard not to confirm my own prejudices, but I couldn't see how her stance would help us get out of the mess we're in. "If you believe all this, why do you keep going?" I eventually asked. This time, there was a moment of flustered confusion, and then she laughed again. "Because I'm irrational!" she exclaimed. "There's never *nothing* you can do."

We parted with an unexpected hug. The darkness may seem all-encompassing. But a glimmer of brightness, it seems, persists in us all.

Clare and I saw things very differently. But I thought we might perhaps both find resonance in the words of the Italian political theorist Antonio Gramsci, who had been imprisoned by Mussolini in 1926 for his role in the Communist Party. He suffered appallingly; as well as the obvious privations, both his physical and mental health collapsed. But as he languished, half-starved and with his teeth falling out, he wrote thousands of pages of highly influential political analysis and commentary, including, in a letter to his sister-in-law: "You must realize that I am far from feeling beaten . . . a man out to be deeply convinced that the source of his own moral force is in himself . . . never falls into those vulgar, banal moods, pessimism and optimism. My own state of mind synthesises these two feelings and transcends them: my mind is pessimistic, but my will is optimistic. Whatever the situation, I imagine the worst that could happen in order to summon up all my reserves and will power to overcome every obstacle."

Pessimism of the intellect, optimism of the will. The ability to see the world for what it is and press on anyway. That seems like a skill worth cultivating. But *can* we cultivate it?

3

Best Possible Selves

Can we learn to look on the bright side?

What actually *is* optimism? One approach is to consider it as a psychological *trait*—an individual personality characteristic that affects an individual's behaviour, doesn't change much over time and is fairly consistent in most situations. All these are true of dispositional optimism, the generalised tendency to look on the bright side. Different people may have different levels of a trait: some people are more extrovert than others, for example. And some people are more optimistic than others. If optimism is a good thing, we might hope that we can increase our levels of it.

The received wisdom used to be that traits were largely settled by the time we reach adulthood, and that's largely but not entirely true: much therapy aims to move the needle on traits. And the line between traits and more transitory "states"—feeling nervous before a job interview, say, and elated afterwards—can be blurry. So maybe it's possible to change our levels of optimism, at least for a while. But to do so, it would be helpful to understand how those levels are set in the first place.

When it comes to the scientific study of any trait, the inevitable question is whether we are born or made that way. And the inevitable answer is "both." We're shaped by our genes *and* by our environments, and by the phenomenally complex interactions of the two. Unpacking these factors can be a formidable challenge. Height, for example, might seem a straightforward vital statistic, but it wasn't until 2022 that we had a full list of the twelve *thousand* genetic variants that influence it—and we're only just starting to understand how they're activated or deactivated by, say, our diet or childhood infections.

Scientists often investigate these questions by studying twins. Identical twins have (almost) exactly the same genes, while non-identical twins have the same degree of genetic difference as any other siblings. Comparing identical and non-identical twins therefore allows us to tease out the contribution that genetics makes to a trait. Comparing twins who are raised together with those raised apart (because they were independently adopted, for example), further allows us to investigate the effects of upbringing.

One of the earliest twin studies to look at optimism was published in 1991 by a team led by the American psychologist Robert Plomin. It looked at more than 500 pairs of twins, half raised together and half apart, and found that around a quarter of the variability in levels of optimism could be explained through their genes. A larger twin study, reported in 2015 by Timothy Bates of the University of Edinburgh, concluded that optimism is indeed heritable but that there are "significant (and substantial) effects of family-level environment and of personal or unique environmental influences."* So what happens to us after we're born also determines how optimistic we are. That is mostly about learning *not* to expect the best, because we are all born optimists. In fact, more than that: we're born *hyperoptimists*.

* Bates's study also added weight to a body of research which suggests that pessimism is, in biological terms, a separate, distinct trait, with overlapping but different genetic and neurological underpinnings to optimism. This is pretty confusing, given our everyday understanding of the term, but might help explain how we can be optimistic about some things while being pessimistic about others. As usual with the human mind, it's complicated.

As anyone who has ever met a young child will know, they look to the future with the very rosiest of expectations and give scarcely any mind to potential obstacles. That intuition is backed up by research using child-friendly tests of optimism, including a junior edition of the Life Orientation Test. Little kids expect that they will become strong, knowledgeable and competent. They assume their weaknesses will over time transform into great strengths—from clumsiness to athleticism, messiness to neatness, foolishness to wisdom. They may even believe it's possible to overcome such deficits as poor eyesight or a missing finger. They also believe the same is true for their friends. In Garrison Keillor's fictional Lake Wobegon, "All the children are above average." Children seem to believe that's literally true.

This isn't hard to explain. Since children's skills and abilities aren't fully developed, they're generally correct to assume that these will improve greatly in future. As to how much they will improve *by*, they're usually surrounded by older kids and adults who seem supremely capable—so much so that it may not be obvious that their skills and abilities vary, too, or that they have any limit at all. As the cliché has it, "You can do anything if you put your mind to it." That's something else children are led to believe is literally true.

At a deeper level, optimism may serve to further youngsters' learning. We've already seen that the function of optimism may be to propel us out of inactivity in pursuit of reward, like the mouse deciding whether to leave its cosy nest. You would expect that motivation to be especially necessary for children, for whom perseverance is critically important. If we were easily deterred, we would probably never learn to write our names, tie our shoelaces or ride a bike. So children might be expected to pay much more attention to positive experiences while learning than negative ones—and they do.

A University College London team asked just over a hundred London schoolchildren, in three age groups ranging from eight to seventeen years old, to play a game about a rocket flying from planet to planet. If they mashed the control button rapidly enough in a five-second period, they would get to the next planet, where they would be rewarded with a stash of virtual coins; if not, they would see the reward, but not receive it. Both the number of button presses needed and the number of coins on offer varied randomly, but the kids weren't told that: they were simply asked before each round to predict how many times they would need to press the button and how many coins they would get if they succeeded.

As it turned out, all the children systematically overestimated the number of coins they could expect. But the youngest group—the eight- and nine-year-olds—predicted significantly more coins than the early adolescents (aged twelve and thirteen), who were in turn notably more optimistic than the late adolescents (seventeen and eighteen). While all groups paid attention to how close their guess had been to the actual answer and updated their next guess accordingly, the younger children paid less attention to negative results than the older kids. They gained false confidence from near misses, but didn't make corresponding adjustments when they were way off the mark. The net result was that the youngest children ended up expecting considerably more coins than adolescents.

When we're small, we do indeed learn from our successes and ignore our failures. And we carry on doing so. The defining feature of optimism, in the psychological sense, is that it gives rise to mistaken beliefs about the world. Error management theory predicts that, overall, these "mistaken" beliefs work for us: we benefit from outcomes we couldn't reasonably have anticipated. But nonetheless we'll also encounter some outcomes that go against our expectations. Things will go wrong; our plans will fall apart; we will be disappointed. In fact, this will happen time after time, and sometimes the experience will be deeply painful. One of my own unrealistic expectations was that no one in my immediate family would ever fall seriously ill; I was entirely wrong about that. How did my optimism survive this blunt encounter with reality? How does *anyone's* optimism survive a lifetime of such definitive truths?

The neuroscientist Tali Sharot decided to find out. She and her colleagues asked people to estimate their personal chances of suffering undesirable events, such as a dementia diagnosis. Then the researchers told them what the average likelihoods of those events actually were, and asked them to estimate their personal risk again. What they found was that people made a bigger adjustment to their individual estimates if the new information was positive (if, for example, the average likelihood compared favourably to their original estimate) than if it was negative.

In other words, we pay more attention to information when it holds positive implications for our futures and less when it has negative implications. Keep that up over a lifetime, and it turns into a persistent and wide-ranging optimism bias.

Learning from Electrocution

How does this reinforcement process play out in our everyday lives? Martin Seligman thinks he knows. The so-called "father of positive psychology" has declared that his mission is "to steer psychology away from the darkness and toward light"—to maintain our well-being, not just cure our ills. And optimism is at the centre of his programme. But Seligman's take on it is quite different from the ones we've seen so far.

Optimism, as defined by researchers like Neil Weinstein, Michael Scheier and Charles Carver, comprises the irrationally positive expectations people have for their future lives. Seligman's suggestion is that those expectations are rooted in the *explanations* people devise for their experiences, particularly the negative ones. If you've heard this idea before, that's because Seligman has not been shy about taking his message to the public through a series of bestselling self-help books, including 1991's *Learned Optimism*, in which he argues that changing the way we think about our experiences can help us form brighter expectations of the future: we can teach ourselves to be more optimistic.

Seligman's line of inquiry started in a way few people would automatically associate with optimism. Back in 1967, one of his colleagues at the University of Pennsylvania was trying to understand how dogs learned by training them to associate a sound with a harmless but unpleasant electric shock (much as Pavlov's dogs learned to associate the sound of a bell with being fed). During the training, the dogs were strapped into a rubber hammock; just after they heard the sound, they would receive an electric shock. During the experiment, they were kept in a compartment whose floor would become electrified, but which they could escape by jumping over a low barrier into a second, unelectrified compartment. The idea was that the dogs would attempt to escape the first compartment as soon as the sound was played, even if its floor wasn't yet electrified.

But to the experimenter's frustration, the trained dogs didn't react at all. Instead of trying to flee when the sound chimed, they would lie there and accept the shocks for as long as they went on. According to the "behaviourist" orthodoxy of the time, animals had no internal mental life and were incapable of forming expectations of the future. Once the association between the sound and the shock had been made, it was assumed that the dogs had no choice but to react to it. But they didn't.

Seligman and his colleague Steve Maier came up with a different explanation: the training had taught the dogs that they were helpless to overcome their situation, and so they didn't even try to escape the shocks. That "learned helplessness" struck them as similar to the way that people with depression often believe there's nothing they can do to improve their situation. So they devised a set of further studies to see if they could gain any more insight. This time, some of the dogs could turn off the shocks while strapped into the training hammock, by pressing a panel with their heads. These dogs *did* try to flee when subsequently placed in the compartment with the electrified floor, so the researchers surmised that they had *not* learned helplessness. The limited control they had been given during training seemed to have become a more general expectation that they could exert control over a situation.

The next important question was whether dogs that *had* learned helplessness could later be taught that they were not helpless after all. So in another round of experiments Seligman and Maier dragged these helpless dogs repeatedly off the electrified floor—and eventually the dogs did begin to jump off it of their own accord. Dogs that had learned helplessness could unlearn it, although that took time and persistence. A better alternative, they found, was to train puppies that taking action was *always* preferable: those puppies grew up into dogs that were "immunised" against learned helplessness for the rest of their lives.

At this point, Seligman concluded that he'd learned all he could from shocking dogs, and promptly discontinued the experiments. Now he wanted to know whether the same principles applied to people. And indeed, similar experiments carried out on human beings—using an unpleasant sound, rather than an electric shock—demonstrated a similar pattern. People who couldn't turn the sound off by pressing buttons in a training session made no attempt to escape when exposed to it in a later test; people who *had* been able to turn it off, or hadn't had any previous exposure to it, figured out how to turn it off quite easily. Humans, too, could learn helplessness and control.

But it wasn't that simple. About a third of the people who took part in the experiments never succumbed to helplessness, while about one in ten was helpless from the very beginning and never made any attempt to turn the noise off. Some people bounced back from defeat very quickly; others stayed gloomily helpless for much longer. Some of those who'd learned helplessness adapted quickly to new challenges, while others threw up their hands. And while some blamed themselves for not being smart enough to "beat" the test, others berated the experimentalists for rigging it. The glass of water, again: seeking an explanation rather than accepting the fact of it.

Seligman thought this was a pattern with much broader implications for our lives. "Who survives when his work comes to nothing or when he is rejected by someone he has loved long and deeply?" he wrote in *Learned Optimism*. "Clearly, some people don't prevail; like helpless dogs, they crumple up. And some do prevail . . . they pick themselves up and, with life somewhat poorer, manage to go on and rebuild." But what accounted for the difference between the two? The explanation, he and his colleagues eventually suggested, lay in how we've learned to explain our failures to ourselves.

Seeking Out Explanation

I've succeeded at a lot of things. I've failed at quite a few, too. One of the most annoying, to this day, is that I didn't pass my driving test the first time.

I failed on a technicality—not checking my mirror before making a turn. My inspector, a stiff and bureaucratic chap, noted this carefully. I thought a less pernickety examiner might have let it go. The second time, the examiner was initially much cheerier, but during the course of the test he seemed to become increasingly convinced I was going to crash the car: he even grabbed at the wheel at one point. He seemed very nervous to be a driving examiner, but needless to say I didn't pass that time either. The third time, I finally got a fair-minded and level-headed examiner, and the pass I should have received the first time.

My friend Annie also failed her test twice, but accepts that she didn't deserve to pass; like me, she passed on the third attempt, but only, she says, because the instructor was kind.* Like me, she occasionally bounces her car off parking barriers, which she attributes to her general tendency to get flustered, as evident in other spheres of her life. And although she's been driving for decades, she's consumed with anxiety every time she gets in the driver's seat. In short, Annie is convinced she's a terrible driver—although she actually has much the same record behind the wheel that I do.

In Seligman's terminology, I have an optimistic "explanatory style," while Annie's is pessimistic. Think back to the first part of the definition of optimism: "An inclination to put the most favorable construction upon actions and events." If I fail—as we all do from time to time, sometimes even twice in a row—then the most favourable construction is the one under which the reasons for my failure are circumstantial and unlikely to be repeated. An optimist tends to explain away failures and screw-ups by citing temporary, external factors that have nothing to do with their own personality or abilities. Conversely, they accept success as a result of their own inbuilt efforts and aptitudes.

A pessimist like Annie, on the other hand, views her own setbacks as an enduring consequence of the person she is and the world she lives in: her perceived failures at driving are in line with her perceived failures elsewhere. For her, success is bestowed by others, or by happy accident. When I encounter a new challenge, I'm confident of being able to overcome it; Annie assumes she'll fail before she even begins. In short, she's a pessimist.

A team led by Seligman's collaborator Christopher Peterson fleshed out this idea with a questionnaire somewhat like Scheier and Carver's Life Orientation Test. But unlike the LOT, which asks general questions and doesn't probe for explanations, the Attributional Style Questionnaire (ASQ) presents you with six positive

* I've obscured her identity, for obvious reasons, but also because I can do without listening to her protest that she really is an awful driver.

and six negative scenarios and asks you to vividly imagine that each of them has happened to you. Half of the scenarios relate to achievements ("You become very rich"; "You have been looking for a job unsuccessfully for some time"), and half to relationships ("You meet a friend who compliments you on your appearance"; "A friend comes to you with a problem and you don't try to help").

The ASQ then asks you to write down why such an event might have happened, then to rate it along three "attributional" dimensions: that's to say, it asks if it was caused by something about you (internal), or something about the situation (external); whether the contributing factors are permanent (stable) or transient (unstable); and whether they are characteristic of many situations you encounter (global) or just situations much like this one (specific).* If you tend to claim that the causes of your (mis)fortune are internal, stable and global, you're deemed to have a pessimistic explanatory, or attributional, style. If you claim they're mostly external, unstable and specific, you're considered to have an optimistic style.

So, in the case of an unsuccessful job hunt, you would first write down the single most important reason you might fail to find a job. You might say it's because you panic under pressure, and always have; just like you do on dates, too. That's a pessimistic explanation, one that suggests you can't do anything about the outcome. Or you might say your interviewer took a dislike to you, and you were in any case feeling off-colour that day. That's a more optimistic reading of the situation.

Does an optimistic explanatory style actually pay off? Seligman and another colleague, Peter Schulman, put this idea to the test with a group destined to suffer frequent, perpetual setbacks: people who sell life insurance. "Sales agents repeatedly encounter failure, rejection, and indifference from prospective clients," they wrote

* The ASQ also asks people to rate how important such an event would be to them, in case that made a difference to the attributions people made; but in the event, it didn't.

when reporting a now-classic study in 1986. That makes it the kind of job people do for a few years when they need to make a living, not the kind of job people invest their careers in. Nearly four out of five sales agents quit within three years, the researchers reported.

What about the remaining fifth? How did they keep bouncing back from being sworn at and hung up on time and time again? Seligman and Schulman sent a copy of the ASQ to all 1,100 sales agents serving the Pennsylvania region for the Metropolitan Life Insurance Company. Ninety-four came back from agents whose performance could be tracked. Sure enough, they found that agents whose ASQ score was above the median sold 37 per cent more policies than those below it; the top decile sold 88 per cent more than the bottom decile. They also gave the ASQ to 104 new recruits and monitored their progress for a year: while there was little difference in performance initially, the more optimistic agents began to outperform in the second half of the year. At year's end, more than half the recruits had quit—but of those who remained, more than two-thirds were extreme optimists. Optimistic explanatory style, then, seems to be associated with perseverance and success, just like dispositional optimism.

How about the opposite? Seligman's assertion was that a pessimistic explanatory style isn't only associated with lack of confidence and self-esteem, but also with underperformance at school and work—with one instructive exception: lawyers.

Together with some colleagues, Seligman got the entire entering class of 1987 at Virginia Law School to take the ASQ, then tracked their subsequent progress through their studies. In stark contrast to the pattern we've seen so far, those who scored as pessimists actually got *better* grades and job offers than their more positive classmates. (Perhaps Annie missed her calling; she is not a lawyer.)

Why? Lawyers are professional pessimists. Nobody wants a lawyer whose attitude towards the small print is *Ah, I'm sure it'll be fine*. Everyone wants someone who will look for every conceivable way that things could go wrong. In law, "pessimism is prudence," explained Seligman. "The ability to anticipate the whole range of problems and betrayals that non-lawyers are blind to is highly adaptive for the practicing lawyer who can, by so doing, help his clients defend against these far-fetched eventualities. If you don't have this prudence to begin with, law school will seek to teach it to you."

The problem, says Seligman, is that once you've started to look at things that way, it's hard to stop when you leave the office. "Lawyers who can see clearly how badly things might turn out for their clients can also see clearly how badly things might turn out for themselves," he wrote. Their pessimistic explanatory style informs everything from career disappointments to relationship failures. What makes a lawyer good at their job may lead them to struggle in other parts of their life: lawyers are more than three times as likely to suffer depression than the general population, as well as more prone to alcohol and drug abuse.

The solution, according to Seligman, is to be a professional pessimist and a personal optimist. To achieve this, he suggests a strategy called "disputation," in which you challenge your own pessimistic thoughts when they arise. Lawyers are particularly well-suited to this, he suggests, because marshalling an argument against a particular proposition is precisely what they're trained for. But he's adamant that anyone can—and should—consciously practise their ability to come up with optimistic explanations. To do this, they can use what Seligman calls the ABCDE model.

A is for Adversity. In *Learned Optimism*, Seligman gives the example of someone inclined towards pessimistic explanations who discovers they've lost an expensive earring borrowed from a friend. B is for their Belief that their friend will be justifiably furious with them for this characteristic irresponsibility, possibly to the point of dissolving their friendship. C is for Consequences: feeling sick and stupid, reluctant to admit the loss to the friend. D is the critical Disputation stage: acknowledging that the friend will be disappointed, but also that she'll most likely realise it was an accident, and that it isn't justifiable to label them as irresponsible because of it. And E is for Energisation: dust themselves off and move on.

Seligman suggests starting with one week of disputing attentively: "When you hear the negative beliefs, dispute them. Beat them into the ground," he said. Reader, I tried it. Perhaps I was just too optimistic to begin with, perhaps I had a particularly stress-free week, perhaps I felt too self-conscious to do the exercise properly. But I found that I didn't have all that much to dispute. It wasn't that negative events and pessimistic explanations never came to mind; it's that I found myself disputing them even before I consciously tried to. For this entirely unscientific sample of one, that seems to validate Seligman's basic idea—but it's not so easy to figure out if it really works for people who aren't *already* pretty optimistic.

Seligman's disputation approach is really just a variation on cognitive behavioural therapy, which teaches people to recognise and change patterns of negative thought and behaviour. CBT is very widely used, and can be effective for depression, so it would make sense for it to increase optimism. There's plenty of anecdotal evidence for that, posted all over the positive-psychology internet—but not much formal research to go on. My inexpert suspicion would be that our explanatory styles are set by a highly idiosyncratic combination of genes and experiences. Perhaps constant, reflexive repetition over a lifetime does make them second nature; whether consciously adopting them can achieve the same effect, I can't say.

I'm also uncertain that changing your explanatory style is necessarily a good idea, at least if you aren't struggling. I'm well aware, thanks to my loved ones, that my tendency to externalise my shortcomings, rather than accept my share of the blame, is not my most appealing trait. It seems a short distance from optimising your explanatory style to self-affirming "toxic positivity," about which many words have already been spilt. Failing to see the downside of your actions, refusing to accept the reality of other people's problems, believing that a positive attitude is mandatory: these are all to be avoided.

Disputation is not the only way to boost optimism. Just asking people to think about the future can have a short-lived effect, as can techniques as simple as asking people explicitly to imagine positive future scenarios, or giving them hidden cues such as optimistic words hidden in a jumble of others. There are also drugs that achieve the same effect. L-DOPA, a chemical that increases dopamine function in the brain, has many effects, including making positive future experiences seem more attractive and suppressing the acceptance of negative information.

But longer-lasting increases in optimism have proven more elusive, with researchers trialling a wide variety of techniques, including cognitive behavioural therapy, mindfulness and meditation, and daily visualisations of eagerly anticipated future experiences, as well as more esoteric approaches such as sensory deprivation and even lying on a bed of nails. Although some of these have proven useful in alleviating depression and anxiety, they have mostly been of limited value in boosting optimism among the psychologically healthy. (You may be relieved to hear that the bed of nails proved to be of no use at all.)

A more plausible contender is the "best possible self" (BPS) exercise, invented by Laura King of Southern Methodist University in 2001. It's simple, fairly quick and potentially enjoyable—which makes it popular with life coaches and self-help guides. The BPS exercise is practised in various forms, but the basic idea is that you spend fifteen minutes each day writing about the version of yourself in a future where everything has gone right. All your efforts have paid off and you have accomplished everything you ever wanted to. Then you spend five minutes imagining that future.

Does it work? A bit. Overall, participants in a best-possible-self exercise seem to enjoy a modest increase in positive affect (around

7 per cent) and become slightly more likely to anticipate positive events in future (around 3 per cent). The effect seems to be temporary, melting away within a week or so, although psychologists are now trying to figure out if practising it more routinely can make the gains more permanent.

I do believe it can. I realised I'd inadvertently stumbled on the best-possible-self exercise during my mourning period: the blog posts I wrote daily were exercises in self-definition. (Being a writer by nature made this a more intuitive approach than, say, counselling.) Some were explicit attempts to describe a more positive future, some were attempts to explain what had happened and some were neither. But they were all informed by the desire to work out who I *could* be now that such a large part of my life had suddenly been rendered irrelevant. It certainly felt as if it made a significant difference. It's also something I carried on doing, from time to time, when I found myself at a crossroads, when I needed to remind myself to be optimistic. Perhaps I'm still doing it with this book.

Understanding Our Imagination

You'll have noticed that as we've got further from laboratories and research projects, and closer to the real world and actual lives, the idea of optimism has become more diffuse and less objective. Real-world optimism is tied to personal significance, desires and needs, which are necessarily subjective. Back to the definition again: optimism is "an inclination to put the most favorable construction upon actions and events or to anticipate the best possible outcome." Seligman's thesis on explanatory style addresses the first part. What about the second? What does it mean to anticipate the best possible outcome?

The best-possible-self exercise is a deliberate attempt to do just this in a personal context. It encourages us to conjure up partners, homes, jobs, holidays, outfits, friendships, romances and revenges; adorable first steps, perfect wedding days, tearful funerals—and not as abstractions, like dates or words or symbols, but as places and events with distinct features and textures. This is what neuroscientists call "episodic future thinking," or "foresight"—visions of our personal futures, extrapolated from the present. But our visions of the future aren't limited to our own experiences—or even to things that exist. If I say "walking on the Moon" or "talking horse" or "superhuman robot," an image will pop into your head, even though you've never actually experienced these things. They are purely the products of our imaginations.

So how do we do it? We can't sense the future directly—you can't see tomorrow, or hear or feel it—so the feedstock for our imagination has to come from somewhere else. Or sometime else. "A mind is fundamentally an anticipator, an expectation-generator. It mines the present for clues, which it refines with the help of the materials it has saved from the past, turning them into anticipations of the future," wrote the philosopher of cognition Daniel Dennett in his 1996 book, *Kinds of Minds*. "And then it acts, rationally, on the basis of those hard-won anticipations."

Our brains are constantly building impressions of the present moment, each around two and a half seconds long: moving from one of these to the next is what makes us aware of the passage of time. Every half a minute or so, those recordings are bundled up into an "episodic memory"—the memory of an experience, rather than of a fact or skills we've learned—which is stashed away for the long-term by a pair of components towards the base of our brains. Each of these is shaped a bit like a seahorse, so the sixteenth-century anatomist Giulio Cesare Aranzi called it the "hippocampus," Latin for seahorse.

We know this, as with much of what we know about the brain, because of what happens when things go wrong. In 1953, a twenty-seven-year-old man code-named "Patient H.M." underwent radical brain surgery, including the removal of his hippocampi, in an attempt to treat his epilepsy. It worked—his seizures became controllable—but he also, unexpectedly, lost the ability to form long-term memories. He could remember his life before the surgery, and he could remember what had just happened. But once his experiences dropped out of that thirty-second window, they were gone for ever. From the moment he woke up from surgery, to the moment he died in 2008, Henry Molaison lived permanently in the present moment. For him, the past did not exist.

Nor did the future. This aspect of Molaison's condition was not investigated rigorously, but other people with damaged hippocampi can't look forward to a brighter future—or, in fact, *any* future. If you have a healthy hippocampus, and I ask you to imagine going on holiday to a white sandy beach in a beautiful tropical bay, a vivid image will pop into your mind's eye: maybe the long curve of the water, a rickety fishing boat out at sea, wind rustling through palm trees. But someone with a damaged hippocampus can only muster a vague impression, perhaps of blue and white. That was what Demis Hassabis found when working at University College London with people who suffered from hippocampal amnesia in 2011.* From this and further studies, Hassabis and his colleagues concluded that the same process underpinned episodic memory, imagination *and* foresight.

Hassabis called this process "scene construction": the brain's ability to generate, maintain and visualise a complex spatial context. This, he argued, enables us to draw on our memories to improve our imaginative depictions of the future. The jury remains out on that, but the general principle that memory, foresight and imagination are intertwined is widely accepted. Ask people to imagine the future, or to remember the past, and the same areas of the brain

* Aphantasics—people who have superficially normal brains but can't create mental images, for reasons that aren't well understood—have similar difficulties with memory and imagination.

light up, hippocampus included. In fact, many researchers believe that's the whole point of memory. After all, what's done is done. We can't change the past, so why is it helpful to relive a previous experience? The obvious answer is that it allows us to learn from our experiences—and to predict the future on the basis of them.

There are three points worth noting about how the brain might do this—"might" because like most things about the brain, it's incredibly complicated and we aren't very sure about any of it. The first is that we don't just store an entire memory in our hippocampus as if it were a recording. Instead, its various aspects are "encoded" in the connections between particular groups of neurons; when you decide, consciously or otherwise, to remember something, the hippocampus strengthens those connections so that they will endure. That creates a kind of giant library we draw from as required to reconstruct a memory, to construct a future scene or to imagine a fictitious event. That might seem fairly obvious, but it's worth bearing in mind as a limitation on our ability to imagine things we've never experienced—and of course, the future will be full of those.

Second, memory storage includes the formation of enduring connections between the hippocampus and those parts of the cortex—the grey outer layer of the brain—which process incoming sensory data. When we recall a memory, neurons within the hippocampus fire up all of these connections; so we re-experience the past event, as reconstructed from the concepts stored in our brains. The hippocampus also prioritises some memories, possibly according to their potential significance for the future. That's why one element of a formative experience—the taste of a madeleine, say—can invoke a vivid, all-encompassing memory.

Third, the hippocampus is integral to our ability to mentally map and navigate spaces: it contains "place cells" that fire only at specific locations, and "grid cells" that fire in patterns reflecting structures in the environment. There's some evidence that the hippocampus stores memories in much the same way that it stores this navigational information, and that the way we remember and manipulate our concepts of people, places, spaces and times aren't distinct: instead, they're all mixed up. And whatever our brains are doing, that's certainly how we *talk* about it. Consider: When young children say they want a *long* time to play, they hold their hands out wide. I ask people who call me *out* of working hours to keep it *short*. We put what's done *behind* us and point to the days that lie *ahead*.* Many of the ways we describe time are the same as the ways we describe space: in many respects, the future is better thought of as a *place* than a time. After all, a place is somewhere you will be one day; a time is just a mark on a clock or a calendar.

All of these capabilities are brought to bear when we use our imaginations. We assemble the concepts that seem relevant; we "pre-experience" what a scene or event would be like; and we transport ourselves into that scene, whether one we might encounter in the future, or a completely imaginary one. This ability to transport ourselves to another time and place is called "mental time travel," and it's not some special feat requiring talent or effort: it's entirely routine. Next time you find your mind wandering, try to pay attention to where it takes you. You'll quickly find that while your body has stayed entirely still, your mind has been to dozens of different times and places, ranging as far as your childhood or retirement, and as near as the next item on your to-do list.

The ease with which we can project ourselves into scenes that don't (yet) exist, or which have ceased to exist, is extraordinary. We mentally assemble not only the scene's physical features,

* Most of us do, anyway. The richness of human culture being what it is, there are always some people or groups who do things differently, although the conflation of space and time is widespread.

adjusted for changes over time, but also the intentions and actions of those who inhabit it. And we can construct not just *one* scene, but many variations on a given theme. Of course, the resolution of our mental images varies—the more remote the scene, the fuzzier our (re)construction of it. You can probably imagine waking up in your bedroom tomorrow with considerably more fidelity than you can imagine waking up in a spaceship orbiting Saturn, or in the year 3000.

And you can more easily imagine a positive event than a negative one. Tali Sharot, with colleagues, asked a group of volunteers to imagine a range of events while having their brains scanned. Those who were imagining positive events—winning an award, say—reported vivid and engaging mental images, whereas those imagining negatives—like breaking up with a partner—reported only vague and blurry impressions of what that would be like. They also reported they felt more distanced from the negative events, as though they were outsiders rather than participants. When the researchers asked how distant the events felt, positive events were deemed to be closer than negative ones (or any past event). The researchers also measured the volunteers' levels of optimism using the revised Life Orientation Test: it was those who scored highly on optimism who felt the positive events to be most vivid and immediate.

The scans backed all this up: two areas of the brain—the amygdala and rostral anterior cingulate cortex, thought to be involved in both visual imagery and processing emotions—sparked to life while the volunteers were imagining any of these events; but they lit up more strongly when the event was positive. Those areas don't seem to work the same way in people with clinical depression, who struggle to imagine positive events in their futures and make pessimistic estimates of the odds that they will come to pass. Training them to create more vivid, positive images might provide a new mode of treatment.

Finding a Direction

This heavy-duty neural equipment is all very impressive, but does it work? The evolutionary just-so story is that we developed it so that we would be able to anticipate what's going to happen next—or perhaps to *change* what's going to happen next. We twenty-first-century types are used to the idea that the future is ours for the making. We're told we can be who we want to be, that we can change the world, that the future is full of possibilities. But does it always feel that way?

Psychologists describe our attitudes to the past, present and future as "time perspectives," which capture which direction we look in when we're thinking about the world, and what we see there. One of the most widely used evaluation frameworks was created by the psychologists Philip Zimbardo and John Boyd, who came up with a questionnaire, the "Time Perspective Inventory," designed to evaluate attitudes to past, present and future. Like Martin Seligman's attributional style questionnaire, the TPI seems as much intended for self-help as for clinical practice; you can take it online, and its authors claim that by being mindful of your type, and continually challenging your assumptions, you can gradually adjust your perspective.

The full version of the TPI has fifty-six items, but here's a shorter version, by Jia Wei Zhang, Ryan T. Howell and Tom Bowerman, which has only fifteen and is a reasonable proxy.*

> Read each item and, as honestly as you can, score your answer to the question: "How characteristic or true is this of me?" according to a scale in which 1 is "very untrue," 3 is "neutral" and 5 is "very true."

* Zhang, J.W., Howell, R.T., Bowerman, T. (2013), "Validating a brief measure of the Zimbardo Time Perspective Inventory," *Time & Society*, 22(3), 391–409: https://doi.org/10.1177/0961463X12441174.

	TIME PERSPECTIVE INVENTORY	SCORE
1	I think about the bad things that have happened to me in the past.	
2	Painful past experiences keep being replayed in my mind.	
3	It's hard for me to forget unpleasant images of my youth.	
4	Familiar childhood sights, sounds, smells often bring back a flood of wonderful memories.	
5	Happy memories of good times spring readily to mind.	
6	I enjoy stories about how things used to be in the "good old times."	
7	Life today is too complicated; I would prefer the simpler life of the past.	
8	Since whatever will be will be, it doesn't really matter what I do.	
9	Often luck pays off better than hard work.	
10	I make decisions on the spur of the moment.	
11	Taking risks keeps my life from becoming boring.	
12	It is important to put excitement in my life.	
13	When I want to achieve something, I set goals and consider specific means for reaching those goals.	
14	Meeting tomorrow's deadlines and doing other necessary work comes before tonight's play.	
15	I complete projects on time by making steady progress.	

The TPI provides scores for five time perspectives; most people don't correspond neatly with one perspective or another, but it's likely that you will score highly on one or two. The first perspective is "past negative," dominated by unpleasant recollections of the past; Zimbardo says people who score highly on past negative tend to be pessimists who "believe the way they view the world is 'the true' reality." Questions 1 to 3 of the short test provide your score on this measure.

Conversely, "past positives" like to reminisce about the good old days and may be small-c conservative and cautious; the score for this perspective is given by questions 4 to 6. "Present hedonists" live for the moment (questions 7 to 9). "Present fatalists," on the other hand, don't expect their actions to make any difference to the world (questions 10 to 12). "Their destiny—and future—is set; they believe they have little or no control over what happens to them and that their actions don't make a difference in the world," writes Zimbardo. "For some, this time perspective comes from their religious orientation, for others, it comes from a realistic assessment of their poverty or living with extreme hardships."

Their diametric opposite are the "future-oriented," who take charge of their own destinies (questions 13 to 15). The future-oriented, Zimbardo writes, "are always thinking ahead. They plan for the future and trust in their decisions; in the extreme they may become workaholics, leaving little time to enjoy or appreciate what they have worked so hard to achieve. But they are most likely to succeed and not get in trouble."

Zimbardo and Boyd don't suggest that any one time perspective is optimal: rather, they appeal for balance, although they do suggest "ideal" scores based on their clinical experience, suggesting that most people should spend less time dwelling on the past and more time having a good time today and achieving their goals tomorrow. Present fatalists, Zimbardo suggests, should spend more time giving themselves selective permission to hedonistically enjoy themselves in the present. So, in fact, should the future-oriented, to make sure they don't get so wrapped up working towards the future that they neglect the pleasures and relationships they have today.

My own scores suggest I'm considerably less fatalistic and prone to mulling over bad experiences than most people; and, unsurprisingly, given that I'm writing these words, that I'm significantly more driven by considerations of the future than the present. (Most of my scores are some way off the "ideal" ones, in varying directions.) I certainly sympathise with the suggestion that future goals can outweigh present concerns. Like many people, I sometimes feel like I'm running hard towards some destination, but never quite getting there.

Whatever the personal relevance of one's TPI, the message that we need to balance consideration of past, present and future is a useful one to bear in mind. But we've already seen that agency, action and self-determination—characteristics most aligned with a future-oriented time perspective—are associated with optimism. Humans aren't content to simply accept our fates: that's not how we've made the journey from roaming the savannah to navigating the urban jungle, from wielding hand-axes to toting smartphones. Now we need to move on to the next stage. And optimism will help us do it.

Towards a Conclusion

We still have a lot left to learn about the psychology of optimism, but we can piece together a comprehensive, if somewhat speculative, story about how it works, from its evolutionary underpinnings to the modern world. That story goes something like this:

We need something to motivate us to act—even when we're uncertain about the outcome, as we are most of the time in the real world. This strategy can work out better than the superficially rational approach of going with our best estimate of the situation, as Martie Haselton and Daniel Nettle showed mathematically. Those of our ancestors who were inclined to be positive about their prospects survived and passed on their genes; over time, we evolved brains that accept information which affirms our positive expectations more readily than contradictory information, as Tali Sharot's experiments demonstrated. That means that we routinely over-estimate our chances of success and expect positive outcomes more often than is "realistic." We're innately optimistic.

As Michael Scheier and Charles Carver showed, highly optimistic people tend to be good at coping with stressful problems, perhaps because their propensity to downplay negatives means they aren't as easily deterred by setbacks or even outright failures. That may be the same reason that they tend to be persistent and resourceful in their pursuit of a goal. That's certainly what Martin Seligman would say.

Suzanne Segerstrom argues that this persistence—at school, at work and in their social and personal lives—in turn means that optimists accumulate the material and social resources needed to flourish: money, friendship, skill and social status. All of these are associated with mental and physical well-being; and optimism might help us to convince ourselves and others of our aptitudes and resources. Having money in the bank and a strong support network makes it easier to envisage a positive future and have confidence that we can overcome problems. So a virtuous cycle begins: more resources leads to more optimism, which in turn leads to more resources.

Optimism, irrational though it might be, is central to the human psyche: it seems to give us an advantage in both everyday life and the evolutionary race. It's closely related to our ability to visualise a different future for ourselves, and choose the best possible one—although we might have to flex our imaginative muscles to do that.

But while we have a finely evolved system for imagining our own futures—and one that for most of us, most of the time, defaults to the positive—we don't seem to have the same inbuilt expectation that our *collective* future will be positive in the same way. In fact, right now, it seems we expect quite the opposite. If we want to improve on that, we need to move away from instinct and towards intellect—to ask what we can expect from the world, and society, that we live in. And that takes us back to where the concept of optimism actually started.

PART II

Optimism Without: Possible Worlds

How can the world be good
when it contains so much that's bad?

Does history follow a pattern,
and what if it doesn't?

Can we see into the future—
and can we control it?

4

The Problem of Evil

How can the world be good when it contains so much that's bad?

Eighteenth-century Lisbon, enriched by centuries of conquest and colonialism, was one of the world's great cities: the heart of an empire that sprawled from the prolific goldmines of Minas Gerais in Brazil to the spice-trading hub of Goa in India and vast agricultural estates in what is now Mozambique. Set around a harbour formed by the Tagus river, the point of departure for many of Portugal's maritime explorers, Lisbon's cityscape was defined by majestic formal architecture standing above a maze of medieval streets: the royal palace, the grand library and, above all else, the mighty cathedral of Santa Maria Maior.

Much of the city's 120,000-strong population was gathered in the cathedral and the city's many churches on the morning of Saturday 1 November 1755—All Saints' Day, one of the holiest occasions in the Catholic calendar. But at 9:40 that morning, the city was struck by a devastating seismic shock. Two more followed in less than ten minutes, mounting in strength; the third was one of the strongest earthquakes ever recorded in Europe. The city could not withstand this pummelling. Its stones crumbled, its streets cracked, its buildings fell.

Terrified Lisboetas fled to the comparatively open areas near the docks on the Tagus, where they were astonished to see that the water had rushed out, exposing the carcasses of sunken ships on the muddy river bottom. But some twenty-five minutes after the quake, the water came rushing back—in the form of a nine-metre tidal wave that swept over the city so quickly that even those on horseback struggled to outrun it. Two more giant waves followed, flooding many low-lying parts of the city.

But the catastrophe was still not over. Toppled candles from the religious ceremonies set light to festive flowers and decorations, quickly turning into fires all over the city. The resulting conflagration blazed for a full five days. When it was finally over, much of the city had been razed to the ground. Perhaps 12,000 of its people were dead; 20,000 more had fled. The palace and the cathedral, seats of regal and religious power, were both in ruins. The consequences were felt far beyond the quake-stricken capital: the entire country's economy was devastated.

In fact, people across the whole of Europe struggled to make sense of this calamity. How were they to come to terms with a catastrophe of this scale, on one of the holiest days of the year? Lisbon's churches were close to the water, while its bars and fleshpots were higher and further from the river's mouth. That meant its most pious citizens had been overcome by earth, water and fire, while its rogues and reprobates remained untouched. Catholics blamed the decadence of the Portuguese court; Protestants blamed the decadence of the Catholic faith. Both factions asked: How could God have permitted such an evil to occur?

The shaken citizens of Lisbon were not, of course, the first to ask themselves such a question. In the philosophy of religion, this is called "the problem of evil," and the West's foundational answer to it dates back to 700 BC, when the ancient Greek poet Hesiod first wrote down the story of Pandora. The most familiar version of the myth is this: Zeus, the supreme deity, angered by the idea that humans are getting above their station, orders the creation of Pandora, the first woman, who is delivered to Earth along with a sealed box which she is ordered not to open. But Pandora cannot resist, and the evils the box contains fly out to assail humanity. She slams the lid shut just in time to trap the last of its contents: hope.

That's how the story goes. But what does it actually mean? The emphasis is usually placed on the release of evil into the world—an explanation for the troubles that have plagued humanity since time immemorial. The more puzzling part, though, is the capturing of hope. What is hope doing in a collection of evils in the first place? Was it included as a sweetener, to offer us comfort and inspiration? Or does it actually *belong* with the evils because of its potential to distract us from the reality of our situations—what we would now call "false hope"? What does it mean that it remains trapped? Has it been kept *for* humanity, or *away* from humanity?

The confusion is partly down to a couple of details lost in translation. In Hesiod's original the container was not a box (*pyxis*) but a large jar (*pithos*) of the type used for provisions—which does make it seem as though hope is being stored for later use. The other, subtler issue is that *elpis* does not necessarily mean "hope." The ancient Greeks used it in both positive and negative contexts, and so some scholars have argued for other translations, notably the more neutral "expectation."* That makes better sense of the Pandora myth: what humanity retained was not so much a positive outlook as the ability to form expectations of the future, rather than simply being buffeted by events as they came. Indeed, Pandora was made by the demigod Prometheus, whose name means "foresight."

Optimism—or for that matter pessimism, realism or any of the other points on the spectrum—is a statement of what we expect of the world. It's part of our worldview. And just as our individual

* The same used to be true of English, but somewhere along the line hope came to mean a positive expectation by default. It is still somewhat ambivalent in Spanish and Portuguese.

worldviews are shaped by our upbringing, education and personalities, so our collective worldviews—as families, groups, societies, nations—are shaped by our histories, knowledge and culture. All of us will experience a moment, sooner or later, that forces us to reconsider our expectations of the world as fundamentally benevolent, cruel or indifferent; amenable to our strivings, or inimical to them. For some, that's a moment to embrace religion; for others, to renounce it.* I think societies have those moments, too. The worldview of the Pandora myth suggests that powerful but fickle gods dictate what happens; the one prevalent in eighteenth-century Lisbon was of a divinity who punishes the wicked but protects the innocent. The 1755 earthquake challenged that worldview.

What we didn't know in Hesiod's day, or in the time of the Lisbon earthquake—and don't really know now, for all that we might like to think otherwise—is whether there is any basis for our expectations to be positive or negative, or for that matter strictly neutral. Is the world constructed, or arranged, in such a way that we can expect it to be generally benevolent to us? Or does it proceed haphazardly, with no discernible purpose? Several centuries after Hesiod there came a more nuanced statement of the problem, attributed to the philosopher Epicurus: "Is God willing to prevent evil, but not able? Then he is not omnipotent. Is he able, but not willing? Then he is malevolent. Is he both able and willing? Then from whence comes evil?" Epicurus's answer was simple: God doesn't care, or even know, about human travails.

Such indifference was not an acceptable explanation for the Christians who came along another few centuries later. An all-powerful and all-loving God was central to the Christian worldview: hope was one of the Catholic Church's three theological

* Why do we have worldviews at all? One suggestion is that they give us a grander perspective that allows us to transcend our fear of death: this is called "terror management theory." So the first part of this book was about error management; this one is about terror management.

virtues.* That made the problem of evil that much more of a problem—particularly (to skip lightly over many centuries of scriptural wrangling) as religion began to cede its explanatory power to a new way of understanding the world and its vicissitudes: science.

The problem of evil is a cosmological version of the half-glass of water. Just as we are not content to accept that the glass is simply as it is, but seek explanations for how it got that way, so we are not content to take the world as it is, but seek explanations for its apparent flaws and imperfections—and what we might therefore expect from it in future. By the time of the Lisbon earthquake, René Descartes and Baruch Spinoza had opened philosophy up to the idea that truth need not solely be "revealed" by close scrutiny of scripture, but could also be deduced through the use of reason: "I think therefore I am" and all that followed. Meanwhile, Francis Bacon, Galileo Galilei, Isaac Newton and others were uncovering universal principles behind motion, gravity, light and much else. Could the problem of evil be addressed similarly, through reasoned argument? Or were God's motives and means utterly beyond human understanding? This was a debate that had already been running for more than half a century—and which had given rise to the philosophical concept that came to be known as "optimism."

Compossibility

Sophie Charlotte, Duchess of Hanover, did not get on with her father-in-law. She had been raised in a family where arts and learning were celebrated, but married into one that prized frugality and martial discipline. Frederick William of Brandenburg-Prussia ruled

* Even in our more secular times, the word "hope" retains a spiritual, even transcendent quality, compared with optimism's mundanity. Hope began as an article of faith, optimism as an exercise in reason: that's why this book is about optimism, not hope.

over his country, and his family, with an iron fist: Sophie considered him, with considerable justification, to be a boor and a brute. The dislike was mutual: Frederick William thought of her as a flibbertigibbet who would, if unchecked, lure his son and heir away from the proper exercise of power, ruining what he'd worked so hard to build. He saw to it that their base in Berlin had absolutely no place for intellectual diversions or frivolous entertainments.

Sophie, shunned as an outcast at court, endured this grinding milieu for four years, until Frederick William died of dropsy in 1688 and Frederick junior ascended to the throne. Sophie, elevated to queenly status, wasted no time refashioning her summer palace, a few miles west of Berlin, into a place of culture and learning. There, she sought to emulate the glorious court of Louis XIV at Versailles, which had inspired her Hanoverian family, tapping into the Prussian state's replete coffers to recruit artists, performers and scholars who would adorn the palace, entertain in its salons and debate the issues of the day in its gardens.

One of her favourite conversation partners was Gottfried Wilhelm Leibniz, whose polymathic genius and meandering career meant he knew something about everything and quite a lot about most things. By the time Leibniz had turned thirty, he had written a thesis about computing, assembled a mechanical calculator, trained in philosophy and law, worked as a diplomat in Paris and London, and contributed to dozens of other fields, ranging from geometry to music theory. He had plenty of admirers, but adversaries, too. Leibniz had been the first to publish a theory of calculus, the branch of mathematics that deals with continually changing quantities, but supporters of Isaac Newton had accused him of taking inspiration from the Englishman's earlier, unpublished work on the same subject.

Having failed to secure suitable employment in one of Europe's statelier capitals, Leibniz had settled for the nominal role of court historian to Sophie Charlotte's family in Hanover. In practice, though, he advised on matters as disparate as mining engineering and the reunification of the Protestant and Catholic churches. Over time, he became an increasingly frequent visitor to Berlin, ostensibly to set up a new Academy of Sciences. During these visits, he and Sophie enjoyed deep conversations on many subjects in the palace gardens. "Madam, there is no way to satisfy you," Leibniz is said to have remarked: "you want to know the reason for the reason."

In the early 1700s, what Sophie Charlotte wanted "the reason for the reason" for was the problem of evil, which had become a hot topic in European salons. In 1697 the scholar Pierre Bayle had poured scorn on pretty much every previous attempt to resolve the problem in his eccentric but influential *Historical and Critical Dictionary*, concluding there was no way to figure out God's motives. Score one to revelation. If reason were to catch up, there needed to be some universal principle that could explain the existence of suffering.

Fortunately Leibniz thought he'd found one. In fact, he'd already been working on it for two decades.

One much-debated question of the then-new science of optics was how light found its way through lenses and bounced off mirrors. The laws of refraction and reflection had been posited, but it wasn't altogether clear why they worked. In a dazzling 1682 paper, Leibniz showed they could be derived from a simple principle: light always followed the "easiest," or "optimal," path from source to destination, no matter what prisms or mirrors stood in its way.* From there, using his newly invented infinitesimal calculus, he demonstrated that a similar process of "optimisation" could be used to evaluate all sorts of other problems: for example, to find the shape of the path taken by a bead sliding down a curved surface under gravity;

* How light "knows" which is the easiest path *before* it sets off is a problem that had to wait longer for a solution: one answer is to think of light as an extended wave, rather than a discrete particle, so the first part of the wave to encounter the mirror or prism changes direction; the rest of the wave then follows.

or to calculate the forces needed to break a solid beam from different directions.

Leibniz's big leap, however, was to apply this "principle of optimality" to the entire cosmos, as the contemporary philosopher Jeffrey McDonough has described. Earlier thinkers, notably Spinoza, had argued that ours must be the only world that could exist, and it must contain everything that could possibly exist. Leibniz, on the other hand, argued that if that were the case, there would be no meaningful role for God to play. He argued instead that God *could* have created the world using many different ingredients, interacting with each other under different laws, and arranged according to different physical and moral schemes. In our world, for example, the Earth revolves around the Sun, but in another, under different laws, the Sun might revolve around the Earth. In our world, people have one head and two arms; in another, they might have two heads and one arm apiece. In our world, people are endowed with free will and brave the consequences of their actions; in another, they might have none, subject instead to lives of perfect boredom or endless torment.

There are thus many *possible* worlds besides the *actual* world we live in. But Leibniz argued that not *all* worlds are possible—only those made up from ingredients and laws that are "compossible." Precisely what he meant by "compossible" has been debated at length by philosophers, but it isn't too far off the idea of "logically consistent." For Leibniz, reason reigned supreme: even God can't put together the elements of the world in ways that defy the dictates of logic or mathematics.

Why not? Here's an example. One of Sophie Charlotte's farthest-flung dominions was the city-state of Königsberg, a port city on the Baltic Sea whose arterial River Pregolya flows around two large islands. These islands were connected to the mainland, and to each other, by a total of seven bridges. Königsbergers would amuse themselves by trying to find a route for their Sunday walks which involved crossing every bridge once, and *only* once. No one had ever managed this low-stakes feat, but no one had ever proved that it was impossible either—until 1735, that is, when the Swiss mathematician Leonhard Euler realised that what was needed was a new "geometry of position," proposed and partially developed more than half a century earlier by—you guessed it—Gottfried Leibniz.

Euler devised a way of writing down the journeys between land-masses as short sequences of letters, then showed that it was impossible to write down a sequence that represented a solution. The details of his solution aren't important for our purposes: what matters is that it shows the Seven Bridges problem to be utterly insoluble.* As the philosopher of mathematics James Franklin puts it: "Although God could make bridges, islands or citizens differently, he could not make them the same while at the same time making it possible for the citizens to walk over all the bridges once and once only."

The broader lesson, in Leibnizian terms, is that the world can only be arranged in particular ways. Leibniz suggested that God, in His wisdom, had chosen the particular arrangement "which is at the same time the simplest in hypotheses and richest in phenomena, as might be the case with a geometric line whose construction was easy but whose properties and effects would be extremely remarkable and of great significance." In short, the arrangement Leibniz describes is a vastly more complex, cosmological equivalent to the line traced by a beam of light.

This would be "the best of all possible worlds," a phrase that has resounded through the centuries—though not, as we'll see, in the way Leibniz might have wanted. For him, it meant that while there might be many possible ways to make a world, there's only one *optimal* way. God has seen to it that ours is the optimal world. And those who take this view came to be known as *optimists*.

* Today's Kaliningrad is a heavily fortified Russian enclave sandwiched between Poland and Lithuania; and the problem has been "solved" by blowing up two bridges.

Theodicy

Was Sophie Charlotte persuaded to become an optimist? While we have Leibniz's testimony that she made vital contributions to his solution to the problem of evil, we don't know for sure, because her husband, Frederick, ordered her papers to be burned after Sophie's abrupt death from pneumonia in 1705. (He'd feared they contained unflattering notes about him.) Presumably Leibniz was pleased, however, because his solution—or their solution—eventually surfaced in the only book he published during his lifetime.

Theodicy was issued in 1710, its title a coinage of Leibniz's from the Greek words for "God" and "justice"—that is, a defence of God's justice, despite the existence of evil—and it sat neatly on the fence between reason and revelation. On one hand, it drew on Leibniz's mathematical and physical research; on the other, it still called upon an all-knowing, all-powerful and all-loving God to ensure that optimality—the best of all possible worlds—was achieved. It was a densely argued work that brought Leibniz's lifelong work in metaphysics to a culmination. It was a theory of everything.

Not everyone saw it that way.

The claim that ours is the best of all possible worlds struck many of Leibniz's contemporaries in the same way it probably strikes you now. This world—this world so self-evidently full of suffering, struggle and strife—*this* world is as good as it gets? Leibniz had an answer to that: even the best of all possible worlds wasn't a *perfect* world. Even in this optimal world there would necessarily be uglinesses that even God could not correct, because to do so would throw other elements out of their "compossibility."

Imagine holding a chain up at both ends, allowing it to sag freely in the middle. It adopts a particular shape—a "catenary"—which isn't one of the familiar family of ellipses, parabolas and so on, but *is* the state in which the forces of tension holding the chain up exactly balance the force of gravity pulling it down, as Liebniz's calculus shows. If you pull down at some point, the relaxed chain becomes a taut triangle: the link you pull down has less potential energy, but links elsewhere go *up* and thus have more.

So, too, the world, which optimises for "goodness" in the same way the chain optimises for energy. If you intervene, to fix one of the problems that we humans are aware of, you might throw out the balance elsewhere. As Jeffrey McDonough writes, "Judas would have been better if he hadn't betrayed Christ . . . But the world as a whole would have been worse, just as pulling down on the middle link of the chain would decrease the potential energy of that middle link but only at the cost of increasing the potential energy of the other links." Leibniz's argument was that whatever evil we perceive here and now, it serves a purpose, or will be compensated for elsewhere, whether in this earthly realm, or somewhere beyond it; it might be in the past, the present or the future.

From our limited perspective—the perspective of those created, the *creaturely* perspective—we can't make out the grand scheme of things. That *cosmic* perspective, evaluating the world in its entirety, is reserved to the creator. By the same token, we can't come up with perfect solutions to our problems. Any attempt to solve a social ill or prevent a disaster may have unintended consequences—perhaps trivial, but perhaps so great as to outweigh the initial problem. That doesn't mean we should do nothing, but we must recognise we can only do the best we can with the information available to our limited creaturely perspective. Only God knows how those efforts fit into the grand plan.

Metaphysico-theologico-cosmolo-nigology

Despite Sophie Charlotte's efforts, her palace remained far from the intellectual centres of eighteenth-century Europe. For Leibniz's idea to make its mark, it would have to be well received in either

London or Paris. But in 1710, the year of *Theodicy*'s publication, Leibniz's long-simmering dispute with Newton finally boiled over after fresh accusations of plagiarism were thrown into the pot. Newton stage-managed an inquiry into the affair by the Royal Society, where he served as president; unsurprisingly, the Society came down on his side, tarnishing Leibniz's reputation.* Leibniz died just six years later, having failed to secure a new patron after Sophie Charlotte's own death, leaving no one to defend his ideas.

To add insult to intellectual injury, a series of English theodicists began stealing the limelight with arguments that were, frankly, crude by comparison to Leibniz's. Most notably, Alexander Pope suggested in his 1733 epic poem, *An Essay on Man*, that rather than being so arrogant as to ask impudent questions and thus disrupt "the great chain of being" laid down by God, we should simply accept that "Whatever is, is Right." That non sequitur is pretty much the entirety of his argument, but it became synonymous with the "best of all possible worlds" argument in many minds. Even today there are people who've heard of Leibnizian optimism but don't realise that there's any more to it than a simple statement of divine righteousness. (Pope himself claimed never to have read a word of Leibniz.)

Things did not go that much better for Leibniz in France. In 1737, a review of *Theodicy* appeared in the influential *Journal de Trévoux*, in which an anonymous Jesuit praised Leibniz's accomplishments, but bluntly dismissed his attempt to figure out God's ways—a straightforward clash of worldviews. It did, however, include this sentence: "En termes de l'art, il l'appelle la raison du meilleur ou plus savamment encore, et Théologiquement autant que Géométriquement, le systême de l'Optimum, ou l'Optimisme":

* The consensus today is that the two men developed their ideas independently: given Leibniz's other achievements, that's perfectly plausible. Nonetheless, sniping continues between rival factions even now.

"In terms of art, he calls it 'the reason for the best' or, even more skilfully, and theologically as well as geometrically, the system of Optimum, or Optimism."

This makes it all the more ironic that the person who eventually brought Leibniz's theodicy back into the spotlight—but caused it to lose what little intellectual respectability it retained—was French, an arch-rationalist and a champion of Newton: the writer Voltaire.

Voltaire had become stupendously famous through his scabrous writings on French high society, lampooning the aristocracy and the Church, and stupendously wealthy through a scheme which exploited a poorly designed state lottery. During one of his many run-ins with the French authorities during the late-1720s, he had been exiled to London, where he learned of Newton's works and became an enthusiastic and eloquent advocate: it was Voltaire who popularised the story that Newton's theory of gravity was inspired by a falling apple.

Voltaire in 1755 was more successful than ever, but he was bitter and tired. He had become estranged from his romantic and intellectual partner, the remarkable natural philosopher and mathematician Émilie du Châtelet, in part because of her interest in Leibniz's ideas about optimality. Without her, he remained a champion of the Enlightenment values of rationality, humanism and liberty, but was unable to make scientific contributions of his own. He had spent decades shuttling around Europe in search of preferment at one court or another. The Lisbon earthquake gave him an opportunity to vent his frustrations.

"We are nothing more than cogs which serve to keep the great machine in motion," Voltaire wrote in the preface to his 180-line *Poème sur le désastre de Lisbonne*. "We are no more precious in the eyes of God than the animals by which we are devoured." The poem itself spoke mournfully of the devastation of Lisbon, taking a swipe in passing at Leibniz and challenging optimistic philosophers to defend their claim that those who had lost their lives and livelihoods were playing their parts in a grand plan for the greater good—if they dared.

A few did. The philosopher Jean-Jacques Rousseau wrote a widely circulated letter to Voltaire the following year, suggesting that the Lisboetas had brought doom upon themselves by overcrowding their city and rushing to retrieve their possessions. Voltaire, never one to back down from a fight, upped the ante by writing an entire book about the foolishness of optimism. *Candide, ou l'Optimisme,* published in 1759, is a satirical romp whose naive protagonist, Candide, is tutored by Doctor Pangloss, an indefatigably upbeat professor of "metaphysico-theologico-cosmolo-nigology." Pangloss continues to insist that "all is for the best, in this best of all possible worlds"—even while he and his companions endure unending calamities: the Lisbon earthquake, shipwreck, impoverishment, war, disease and mutilation.

This belief leads Pangloss to perpetually come up with convoluted rationales as to why things are not as dire as they seem. After catching syphilis from a chambermaid, for example, Pangloss traces it all the way back to Columbus's voyage to the Americas. Syphilis, he reasons, is a worthwhile price to pay for access to such delicacies as chocolate. He also believes that there is no need to avert or remedy misfortune or injustice, because there will be a greater good to follow. When Jacques, Candide's benefactor, is drowning in the bay of Lisbon, Pangloss merely looks on, convinced that this must all be fine. Under Panglossian optimism—Panglossianism, for our purposes—there's no point in even *looking* for ways to make the world better than it is.

Pangloss is very obviously an acerbic caricature of Gottfried Leibniz, who by this time had been dead for more than forty years. Pangloss's arguments, too, are at best a caricature of Leibniz's. Leibniz scholar Lloyd Strickland argues that the German never sought to minimise the real suffering of people caught up by "evil" and never suggested that we should simply stand by and accept it;

instead, he was clear that we should act to prevent suffering wherever we could, since our own actions are inherent to the cosmic plan. Nor does Voltaire actually refute Leibniz's argument or put forward any coherent alternative: he just pours scorn on it. In fact, his ire might have been better directed at the cruder optimisms of the English school, with which he was familiar. Why he picked on Leibniz so mercilessly is unclear. Perhaps it's a back-handed tribute to how memorable Leibniz's "best of all possible worlds" formulation was. Maybe it was related to his adoration of Newton, whom he revered as almost a god among men.* Or it might have been prompted by the memory of his "betrayal" by Émilie du Châtelet.

Regardless, the damage was done. *Candide* proved enormously successful. Published in five countries more or less simultaneously, it's believed to have sold between twenty and thirty thousand copies in its first year alone, and remains one of Voltaire's most celebrated works today. Through its translation, "optimism" finally entered the English language, as foolish positivity that flies in the face of all evidence—an association that has persisted even as its original formulation has been largely forgotten (although not by the dictionary: "the best of all possible worlds" is actually the *first* definition Merriam-Webster gives).† When I say "the best of all possible worlds" today, almost everyone who recognises the allusion cites *Candide*—not Leibniz.

Voltaire took a more direct stand against optimism some years later when he issued his *Pocket Philosophical Dictionary*, in which he slated not just Leibniz's solution to the problem of evil ("He wrote fat books which he didn't understand himself") but pretty much

* Voltaire's 1752 short story "Micromégas" tells of a giant, eight leagues tall and similarly endowed with intellect, who travels from the star Sirius to visit Earth. Textual clues suggest strongly that he's meant to be Newton.

† Another character, Martin, concludes after being abused and abandoned by his family that this must be the *worst* of all possible worlds. The *Journal de Trévoux* again coined *le mot juste* in its review: *pessimiste*.

every other solution that had been proposed, too, including the Pandora myth ("purely to get at Prometheus"), the Manicheans, who believed earthly troubles stemmed from a cosmic tussle between good and evil ("no trivial instance of stupidity") and the English school ("their system saps the very foundations of Christianity but explains nothing"). "So far from being consoled by the opinion that this is the best of all possible worlds," Voltaire concludes, "philosophers who embrace it find it depressing."

Candide closes with Pangloss glibly reciting, one last time, how the series of unfortunate events related in the book has all led up to Candide peacefully eating "preserved citrons and pistachio-nuts" in tranquil Constantinople, which Candide brushes off: "Il faut cultiver notre Jardin"—"We must cultivate our garden."

Generations of reviewers have unpicked this line in various ways. One suggestion is that it alludes to the Muslim tradition of meticulously designing gardens so as to create the closest thing to Heaven on Earth: they are in Turkey, after all. It's also been suggested that it was inspired by an earlier visit to Pope's idyllic garden in London, where Voltaire may have developed a conception of what the best of all possible worlds might really be like. Another widely received thesis is that it means you should mind your own affairs as best you can, rather than concerning yourself with global alarums and trying to devise grand theories of how the world works—a kind of modest, domestic version of optimism.

Perhaps the connotations of rootedness and home-making were appealing to a man who had spent most of his life on the run, or on the road, for one reason or another. In the year of *Candide*'s publication, Voltaire acquired a large estate in the remote village of Ferney, "a miserable hamlet" surrounded by swamps, where he lived for the last two decades of his life—by far the longest period he ever spent in one place. True to his word, he not only restored its chateau and its grounds for his own use, but became a local benefactor, building a church, fostering local artisans and founding theatres to attract visitors. Centuries later, they are still coming to the garden Voltaire cultivated.

Cosmology

The Leibniz–Voltaire debate encapsulates our conflicted attitudes to optimism remarkably neatly, given that one of the disputants was dead and the other didn't seem to have engaged with the text he was supposedly debunking. What's mostly come down to us from it is Voltaire's satirical version of Leibniz's optimism: a philosophical schema for understanding the world transformed into a psychological tendency towards foolish self-delusion.

This bastardised view is not without truth: while Pangloss was a wildly exaggerated spoof, our world is full of Panglossians, many of them elevated to positions of power and influence. Just as the sophistries of the fictional Pangloss led the naive Candide astray, so the over-confidence and neglectfulness of real Panglossians can have consequences for all of us. We'll come back to them later. But if we want to look for more useful approaches to optimism, we should go back to the positions marked out by Leibniz and Voltaire.

Leibniz's approach was to look for evidence of a *grand plan* that shaped the whole world. He believed his law of optimality applied to phenomena large and small, from a hanging chain to the entire cosmos. The world is optimal because God, given the choice of creating many possible worlds, chose the best one. This provided an elegant answer to the problem of evil—but one that is critically dependent not on reason, but faith. Take away the divine element and it is no answer at all.

Voltaire, on the other hand, scorned the idea of a grand plan. He saw no reason to accept that this is the best of all possible worlds and observed, through the character of Pangloss, that it can be damaging to act as though it is. For Voltaire, optimism seems to have been more about *gardening*: cultivating that which is under your control. Perhaps many local efforts will add up to a greater whole; perhaps not. But it is the best that we poor creatures can do.

Voltaire also reasoned himself to a standstill. He was so in love with rationality—as a Newtonian intellectual ideal, as personified by Émilie du Châtelet—that, somewhat perversely for a writer, he disregarded the power of belief, imagination and possibility. Since, in his view, reason could not dispel the problem of evil, those debating it were merely "convicts playing with their chains." Voltaire presents a fundamentally fatalist view of the world. It doesn't really matter if it's the best of all worlds, or the worst: there's nothing we can do about it. The world simply is as it is.

Ironically, the space for imagination is opened up by the more mathematical approach taken by Leibniz. The primary motive for his theodicy might no longer speak to our times, but it retains many useful explanatory features. Whether you believe in God or not, it reminds us that we cannot see or understand the whole design of the world. If we take matters into our own hands, our attempts to fix one problem will have consequences elsewhere that we can't anticipate. And there are logical and practical constraints that make perfection unattainable. We shouldn't strive for a *perfect* world; we should strive for the best of all *possible* worlds.

It's this last notion, that of possible worlds, which I think is most far-reaching. Many people prior to Leibniz had written about travellers' journeys to other worlds, be they exotic islands far across the ocean, or mystical realms on some ineffable astral plane—but he was the first to think of our world as just one of many possible combinations of "compossible" ingredients, one of an ensemble of variations on a theme. And here, as on many other occasions, he proved ahead of his time: the idea of many possible worlds has come back into favour.

In the 1960s philosophers started to think again about possible worlds—ways the world *might* have been—in order to deal with some puzzling questions in so-called "modal" logic: What does "necessary" really mean, or "possible"? These could be addressed in terms of the existence of worlds in which a statement was true, and those in which it was false. If something can be true in *at least* one world, it's possible. If it logically has to be true in *all* worlds—two plus two is four, there is no way to cross the seven bridges of Königsberg without duplication—then it's necessary. And if it happens to be true in our world, but didn't have to be, it's "contingent": Leibniz was German, Voltaire was French.

Some philosophers claim there's no reason to think of these possible worlds as any less real than ours; others scoff at the very idea. Most just use it as a way to think about logic problems. But at around the same time, a strikingly similar idea—but this time grounded in verifiable facts and repeatable experiments—was making its way to the fore in one of the most abstruse yet most successful branches of physics: quantum mechanics.

Quantum mechanics is the framework that considers how entities at the atomic and subatomic scales behave. It's been validated by every experimental test that's ever been devised—which is perplexing, since it's notoriously counter-intuitive, often suggesting multiple answers to a question, some of them mutually exclusive. If we try to figure out the location of a subatomic particle such as an electron, for example, there can be several possible answers until we make a measurement to settle the issue, whereupon just one of those possible answers somehow transmutes into a hard fact. This indeterminacy doesn't make a lot of sense to us—in our macroscopic world, things are where they are—but experiment after experiment has shown that it's how the quantum realm works.

No one is entirely sure what to make of this—physicists included. But one interpretation, put forward by Hugh Everett III in 1957, is that what actually happens is that *all* the possibilities occur—but in different universes. These universes share a history up to a point, but then they branch off and evolve independently thereafter. Because there's a huge number of such alternatives at every juncture, vast quantities of new universes are continually being spawned—but we can never reach them.

Initially considered no more than an ingenious answer to the indeterminacy problem, Everett's idea gradually became widely known over the course of the 1970s. Today, it's one of the mainstream interpretations of quantum mechanics. And again, some cosmologists embrace the idea with Leibnizian zeal; others dismiss it as incoherent and unscientific.

Modal logic and quantum physics are nose-bleedingly rarefied subjects—but nonetheless, these concepts may well sound familiar, even if you've never paid the slightest attention to them. (In fact, I'm counting on it, since the description above is just the barest sketch of these subjects.) The idea that there are other worlds—multiverses, parallel worlds, alternate timelines, counterfactual histories—has become deeply embedded in popular culture. We understand, as a fictional conceit if not as a physical fact, that every action we take gives rise to multiple parallel worlds, one for each possible outcome or choice.

Faced with a fork in the path, we can go left or right. Go left, and we end up in one place; right, and we end up in another. Run for a train: if we make it, one chain of events ensues; if we don't, another chain takes its place. A bullet hits, a bullet misses; lovers part or stay together. In this world, I wrote this book and you're reading it; but in another, I spent the time watching *Frasier* re-runs and you're reading *Candide* instead.

These imaginings may be strangers to scientific accuracy, but that doesn't detract from their usefulness as a central metaphor for attempts to imagine a better world. For our purposes, parallel worlds are just a helpful way to think about consequence and happenstance: about the choices we make; the choices other people make for us; and those that an indifferent universe forces upon us.

Alternity

Why does possible-worlds thinking appeal to us so much if we can't even tell whether possible worlds exist? Perhaps because it's the way we already think about the world, and specifically about the future. As we saw earlier, we imagine different scenarios for ourselves by recombining the elements of our lives. Most of the time we do this without really paying much attention to it, but as the best-possible-self exercise suggests, consciously taking control of the process can help us feel more positive about our chances of achieving our goals. We also do it prolifically: we can generate an infinity, or an eternity, of alternatives—an alternity, if you like—without any particular effort.

In fact, the first major expression of alternity in the twentieth century is the product of imagination, not intellect or experiment. Jorge Luis Borges's hugely influential 1941 short work, "The Garden of Forking Paths," describes a novel, written by a retired civil servant named Ts'ui Pên, which includes every possible outcome. What are initially taken to be many drafts of the same chapter turn out to be the intended text of the book. "In all fictional works, each time a man is confronted with several alternatives, he chooses one and eliminates the others," explains the garden's curator, but "in the fiction of Ts'ui Pên, he chooses—simultaneously—all of them. He creates, in this way, diverse futures, diverse times which themselves also proliferate and fork."

The concept of parallel worlds was relatively new when Borges wrote his story: Murray Leinster's short story "Sidewise in Time," published in 1934, describes a collision between different timelines, and established "alternate history" as a subgenre (or perhaps sibling genre) of science fiction. The idea grew steadily more popular, until by the 1960s *Star Trek* was making forays into the "mirror universe," where you can tell everyone is evil because Kirk wears a muscle vest and Spock has a beard. The alternatives became more and more outré: I fell in love with the genre via Harry Harrison's 1984 novel, *West of Eden*, set in a world where the dinosaurs weren't wiped out by a meteorite impact, but developed a sophisticated culture in which bio-engineered animals take the place of machines.

Eventually alternates went mainstream, the Rubicon being crossed in the 1998 romcom *Sliding Doors*, starring Gwyneth Paltrow. For the uninitiated, *Sliding Doors* tells the story of a woman who in one narrative strand catches a Tube train, and in another has its doors slam in her face. From there, she goes on to lead very different lives (for one thing, she has long brown hair in one, and a blonde pixie cut in the other). Since then, there has been a plethora of media in which the protagonist slips into a different world, changes the one we live in through temporal meddling, or simply inhabits a world that's simultaneously familiar and strange.

Critically acclaimed literary fiction now trades routinely in alternate histories, both domestic and grand (Kate Atkinson's *Life After Life*, or Philip Roth's *The Plot Against America*); but so do blockbuster movies: the multiverse is both the default setting for Marvel's endless spandex melodramas and the framing metaphor for the Oscar-winning oddity *Everything Everywhere All at Once*.* Video games provide quick-twitch sensation by jumping between parallel worlds (the Half-Life series); there are online communities dedicated to producing lovingly detailed maps of imaginary countries; memes depicting different branches of the multiverse and "speculative zoology" contests for art depicting the flora and

* The ultimate inspiration for most of these stories is the "many worlds" interpretation of quantum theory; the collection of universes is often referred to as the "multiverse," although that has a specific, and different, meaning in modern cosmology.

fauna of environments that have vanished, yet to come or never existed. TV shows trust their audiences to follow them through multi-season explorations of alternate history (like *For All Mankind*, in which the Soviets land men on the Moon first), or to get the gag when it's a one-off gimmick in sitcoms ranging from *The Simpsons* to *Friends*. Or *Frasier*, one episode of which features Dr. Crane enjoying and suffering two very different Valentine's Days. Perhaps that's what my alternate watched instead of writing this chapter.

In these depictions, unlike in the real world, characters often breach or glimpse other possible worlds past, present and future. Their appeal comes from making us think about other ways the world might be. But entertaining as all of this is, can we use possible-worlds thinking to sharpen our more general expectations of the world beyond our personal lives? I believe we can and do, and this possible-worlds view of optimism is more powerful than the glass-half-full version we're familiar with. Rather than presenting us with a conundrum which has no real answer, possible-worlds thinking invites us to investigate the full range of possibilities and then, to co-opt Leibniz's definition, choose the *best* of those possible worlds to realise.

The key to making Leibniz's version of optimism relevant to our secular, twenty-first century worldview is to make "the best of all possible worlds" a statement of aspiration, not belief. Stripped of its theological motivations and paraphrased in terms we might use today, Leibniz's version of optimism is one in which we all have parts to play in bringing about the best possible world, given the physical, technological, social and practical limitations that constrain our choices. Not a *perfect* world, not a utopia: the best *possible* world.

This might sound abstract or fanciful. But in fact there are a range of ways in which possible-worlds thinking is applied, many of them down to earth and serious. We use possible worlds—scenarios, counterfactuals, simulations, stories—to forecast the weather; to keep banks safe; to test military strategy; to understand and devise responses to disasters, pandemics and crises. Billions of dollars, millions of lives, the trajectory of nations and populations are determined, more or less explicitly, by possible-worlds thinking.

Generally, however, we contemplate only a few fairly conservative possibilities—a narrow set of "more of the same." The challenges of our times may demand a fuller exploration of the options open to us; and while that can be disorienting, it can be stimulating, too. One way to gain comfort with the idea is to look *backwards* through the possible-worlds lens, to the events of the past and how they might have played out differently.

Our present, after all, is just one of the possible worlds of the past. Our today is yesterday's tomorrow; our present is the past's future. And seeing that things didn't have to turn out the way they did today can reassure us that they don't have to turn out any particular way tomorrow either.

5

The Accidents of History

Does history follow a pattern, and what if it doesn't?

Two months after the Lisbon earthquake, a pair of articles appeared in the *Königsberg News and Advertiser* setting out to explain it. But unlike previous attempts, these addressed the earthquake's physics, not its metaphysics. The author of the voluminously titled "Concerning the Causes of the Terrestrial Convulsions on the Occasion of the Disaster which Afflicted the Western Countries of Europe towards the End of Last Year" claimed that the earthquake had been caused by the explosion of a vast build-up of combustible gas deep beneath the Tagus river. There was no mention of God's role whatsoever. While that explanation was incorrect, the attempt marked a newer, more scientific approach: a first stab at seismology.

Both pieces were written by Immanuel Kant, a lifelong resident of Königsberg, home of the seven bridges, and earthquakes were just one of his many scientific interests. In the same year he published a theory that nebulae, vast clouds of gas and dust in space, would collapse under their own gravity to form stars and their families of planets—which is roughly what we still think today. His very first published work was a response to Émilie du Châtelet's 1740 book *Lessons in Physics*. But Kant is primarily known to us today as a philosopher, not a scientist: the thinker who, more than any other, turned philosophy away from "revelation" of God's ways and towards the uses of reason in understanding the world.

Some Kant enthusiasts pinpoint the beginning of this turn to the Lisbon earthquake: in a later monograph on the subject, Kant wrote: "Man is so opinionated that he sees only himself as the object of God's activities." Instead, he argued, the quake should be seen as the product of natural laws, wholly indifferent to human concerns—a very Voltairean, or even Epicurean, perspective. A quarter of a century later, in 1781, he published his *Critique of Pure Reason*, in which he posited that our understanding of reality is constrained by our senses and shaped by our minds.* That means reason alone cannot deliver us truths about the underlying nature of the world, particularly when it comes to things we cannot directly experience—the existence and motivations of God among them. It's from Kant that we get the word *Weltanschauung*: worldview.

Kant went on to apply that argument to the problem of evil, culminating with the publication in 1791 of a short essay with a sweeping title: "On the Miscarriage of All Philosophical Trials of Theodicy." Like Bayle almost a century earlier, he concluded that there is no way to reason our way to a solution of the problem of evil, although his approach is much more abstract and cerebral. But Kant did not close the door on theodicy completely, says contemporary philosopher Mara van der Lugt: instead, he suggests that "the best of all possible worlds" is not the intellectual resolution of a theological paradox through pure reason, but a statement of *practical* reason: deciding how to act upon real challenges and real problems. The upshot is that we should no longer look to God to ensure this is the best of all possible worlds: we should see to that ourselves. It's from Kant that we get the idea of optimism as a moral duty—not inherent in the design of the cosmos, but a quality we must find within ourselves.

* This is very much how we think of our perception of the world today, one upshot being that we're subject to sensory and cognitive illusions, as described in Chapter 1.

Stepping away from a theologically grounded worldview, established over hundreds of years, is no minor challenge. But Kant argued that transformative insights lead to dramatic changes not just in our understanding of how the world works, but also in how we think about our place in it. In the physical realm, Copernicus's realisation that the Earth revolved around the Sun, rather than the other way round, had led to a dramatic reappraisal of our cosmic status (or lack of it). Over the course of the nineteenth century, there was no shortage of such transformative insights into the nature of atoms and elements, electromagnetism, germs and vaccination, heredity and natural selection and more.

The locus of western thought began drifting away from a metaphysical view of the world, in which understanding God's plan was the objective, to the more empirical one Kant had hinted at, in which God was no longer considered to actively intervene in earthly affairs. That didn't happen overnight, of course: philosophers continued to debate the problem of evil, the basis for optimism and the nature of hope. (Some still do, some of them in theological terms.) But for the most part, the debate increasingly revolved around questions of human agency rather than God's purpose. What we can expect from the world, for better or worse, becomes a question for us; the solution to the problem of evil lies in what we can do for ourselves.

The Industrial Revolution provided one answer: we can remake the world. Colossal engineering projects, mass manufacturing, international trade and transport networks—all fuelled, both literally and figuratively, by resource extraction on a vast scale. This might sound triumphalist, but this period saw rapid, spectacular improvement in our standards of living by almost any metric you care to mention. God's favour might be invoked as a contributing factor, as it was in Helen Keller's paean to optimism, but it was human ingenuity and enterprise that took prominence. It increasingly looked like any divine influence was felt only through the operation of natural laws, as Epicurus, later Voltaire, and subsequently Kant had declared; and as though optimism lay in our ability to remake the world for ourselves. But the notion of the best of all possible worlds was not to be forgotten so easily: if it wasn't the world of today, perhaps it would be the world of tomorrow.

The World Spirit

Human beings are wired to look for patterns and purpose. As social primates, we want to know who did what and why they did it, helping us to understand where we stand, who we can trust and who we should fear. As animals gifted with foresight, patterns help us predict what might happen next, identifying regularities we can depend on. These tendencies have served us well through our evolutionary history, but they can lead us astray when no such purpose or pattern exists.

Natural disasters confront us, in the most dramatic possible way, with the realisation that our expectations of order may be misplaced: hence the philosophical mêlée that followed the Lisbon earthquake. We want to restore cause and effect, or a sense of purpose, to the world. The same desire for order is evident when it comes to human affairs. Rather than accept the sound and fury of human existence at face value, generations of thinkers have devised grand theories of how history proceeds according to some covert scheme.

No one epitomises this tendency more than Georg Friedrich Wilhelm Hegel (Wilhelm to his mother). Hegel was a fan of Leibniz and a follower of Kant, neither of whom is the easiest of reads; but his own philosophy is next-level impenetrable. ("Only one man ever understood me, and he didn't understand me either," he supposedly lamented on his deathbed.) Hegel's approach was rooted in a method known as *dialectic*, originally meaning a dialogue which aims to settle differences through reasoned argument. Hegel's version was to put forward contradictory positions—a "thesis" and "antithesis"—and gradually reconcile them to form a new position—"synthesis"—which overcomes the objections to both.

While capable of delivering great insight, dialectic is also inherently tricky to follow. This is particularly true of Hegel's philosophy of history, which he presented in lectures at the University of Berlin between 1822 and 1830. Hegel introduced his lectures as an exercise in theodicy; but in his philosophy, the guiding principle is not optimality, but a *Weltgeist*—"World Spirit"—that pervades and guides human civilisation towards what we might, to avoid several thousand words of explanation, call "enlightenment." To be sure, there might be occasional setbacks and stumbles, but ultimately our ascent would continue. In fact, it was part of the process that a particular situation (or worldview) would inevitably provoke opposition, which would then resolve into a new and improved situation (or worldview): thesis, antithesis, synthesis.

For example, you might argue that the excesses of the French monarchy (thesis) had brought about disgruntlement among the bourgeoisie, culminating in the French Revolution (antithesis). Hegel admired the goals of the Revolution—basing politics on philosophy rather than preferment, and governance on principles of liberty, equality and fraternity, rather than the wielding of power—but recognised that it ultimately failed to achieve those goals. However, he also believed that occasionally an outstanding individual—Julius Caesar, for example, or Alexander the Great—who embodied the *true* spirit of their times would help things along, an idea which would later morph into the "Great Man" theory of history. In the case of the French Revolution, what followed was synthesis via the person of Napoleon Bonaparte, who took on an imperial mien but preserved many features of the Republic.

Hegel's most famous successor, Karl Marx, also used the dialectic method. Marx contended, however, that the *real* force behind human history was not some ineffable spiritual force, but the material conditions of humankind, with the struggle being primarily between social classes. Nonetheless, his work, like Hegel's, is often read as suggesting that history proceeds in a particular direction—in Marx's case, from slavery to feudalism to capitalism, and then to socialism and communism, with technology being the ultimate driver of change. One descendant of that was the idea of "history from below," in which the struggles of ordinary people create irresistible currents of change—just as individual responses to the current state of the world are having major social consequences, as we saw in Chapter 2.

Whether they were directly influenced or similarly motivated, many other historians echoed this emphasis on how the apparent struggles of the world are actually part of a movement towards a better world—perhaps even the best of all possible worlds. Such narratives tend to be selective: "Whig history," popular during the later days of the British Empire, describes a progression from benightedness to enlightenment, stressing the significance of personal liberty, constitutional monarchy and scientific progress. (The roles played by exploited peoples and expropriated resources don't get much scrutiny in such triumphalist narratives: the best of all possible worlds, but only for some.)

Leibniz, Voltaire and other earlier theodicists had been occupied with what Mara van der Lugt calls "value-oriented" optimism, which asks questions about the fundamental nature of the world: Does good outweigh evil? What role do we play in the divine plan? Is it better to exist than to not exist? In this framing, optimism essentially comprises a belief about whether the world was designed by God to favour humankind, or whether it is indifferent or even hostile to our wants and needs.

Now the question was not so much "Why does suffering exist today?" as "Will there be less of it tomorrow?" The preoccupation of the nineteenth century was *future-oriented* optimism, where the questions were about what we can expect of our individual and collective futures: Is progress real, and is it to be expected? How can we do our part to make the world better? Are we justified in bringing new lives into the world? The worldviews of Hegel, Marx, the Whig historians and the rest were built on those of preceding generations, but introduce the ideas of progress and human agency. Optimism—specifically Keller's "optimism without"—starts to look to the potential of tomorrow, rather than the challenges of the past and present.

This shift makes sense when you consider that the notion of a future that is distinctively *different* from the past is relatively modern. For most of human existence there might have been uncertainty in individual human lives, but the backdrop—the environment, the technology, the business of subsistence, the structure of society, the climate—changed only slowly, if at all. It's only natural, in such an environment, to devise cyclical explanations for the world: the chariot of the Sun riding across the sky between dawn and dusk, the seasons coming and going with the movements of the gods. That's not to say the ancients didn't make plans and promises, just as we do. But they didn't expect the shape of life to change dramatically; if it did, the consequences might be dire. Stories that do describe the cycles of life ending—the Norse Ragnarök, the Christian Apocalypse—present it as a catastrophic event, pushed off to some indeterminate time and place.

But since the onset of the Industrial Revolution, we have become increasingly used to the idea that we might grow up in one kind of world and die in another. The order of things is no longer necessary; it becomes contingent on our choices and actions. One way this shows up is in the emergence of great ideologies: the sense that the future is now a space of possibilities naturally leads to ideas about how it might be filled. And one way *that* shows up is in the reinvention of utopia in the late nineteenth century.

People had been writing about perfect societies for millennia, from *The Republic* of Plato through Thomas More's original *Utopia* to Daniel Defoe's *Robinson Crusoe*. But these utopias were almost always isolated *places*, usually far removed from the narrator's home: that meant they could be safely regarded as models, or satires, of their society without implying that any actual disruption was needed to the prevailing social order.

Once the pace of change became disorienting, however, authors started producing fiction about travelling into the future. *The Year 2440: A Dream If Ever There Was One*, written by Louis-Sébastien Mercier in 1770, tells the story of a Frenchman who falls asleep and wakes up seven centuries later to find his country transformed into a secular and democratic (if rather puritanical) paradise after a revolution; the actual French Revolution took place a mere eighteen years later. The book was a huge success, going through more than forty editions and selling tens of thousands of copies.

More than a century later, the same device featured in the American journalist Edward Bellamy's 1888 book *Looking Backward: 2000–1887*, and in the British designer and author William Morris's 1890 *News from Nowhere*. In both books, the protagonist falls asleep and awakens to find their country transformed into a socialist utopia. Their utopias are quite different, however. Bellamy's is an urbanised, technological wonderland in which machines do much of the work; Morris's, written as a rebuttal, is a pastoral fantasia in which people work for the artisanal pleasure of it.

All three books were taken quite seriously as statements of intent, just as their authors intended. The imaginative heft of a novel can travel further than any earnest manifesto; fiction enlivens the presentation of ideas that have not yet come of age, which makes it useful for promoting them at the time, and for understanding them retrospectively. Morris's has aged better: its call for the fusion of craft and labour is still cited today. Bellamy's sold in vast quantities—only outdone by *Uncle Tom's Cabin*—and hundreds of

"Bellamy Clubs" were set up to discuss the book's ideas. Many were eventually implemented, from bank cards to cooperative stores, although you may have noticed that the US has yet to become a socialist paradise.

The idea that there is no guiding hand, leaving the future malleable by our efforts, is somewhat intoxicating. What you lose in terms of God's protection and munificence, you gain in terms of freedom and opportunity. But once you've got used to that idea, another one immediately presents itself. What if something intervenes to make the future *worse*?

The Year Without a Summer

There are no facts in the future. There are certainly things we *think* of as facts: the Sun will rise tomorrow and shine the day long, spring will follow winter, flowers will bloom and leaves fall. But even these apparent verities are all ultimately presumptions, not facts. Sometimes they turn out to be false, to shocking effect.

In April 1815, Mount Tambora on the Indonesian island of Sumbawa exploded with truly cataclysmic force: it remains the largest volcanic eruption in recorded history. The volcano literally blew its top: the cone of its summit was obliterated. Tens of thousands of islanders were killed and made homeless, but the volcano's effects spread much further than that. As well as spewing rock, ash and lava, the volcano threw vast quantities of aerosolised minerals, notably sulphates, into the atmosphere. And sulphates, when they mix with atmospheric gases, reflect incoming sunlight back out into space, meaning that the Earth's surface becomes cooler.

This effect was most pronounced in southeastern Asia, where torrential rain and the reduction in sunlight meant crops failed utterly, but it was felt around the world. The average global temperature may have dropped by a degree, on top of a previous long-term cooling trend. By 1816, the effects were also significant in Europe and some parts of North America. During "the year without a summer," heavy rain and exceptional cold—with frost being reported in June—destroyed crops, leading to widespread food shortages. The sky was often unnaturally dark; but sometimes it glowed balefully, due to volcanic haze, as commemorated by some of J.M.W. Turner's most spectacular paintings.

That was not the only art inspired by the event. The incessant rain confined a party of writers including Lord Byron, Percy Shelley and Mary Shelley to quarters during a Swiss holiday; famously, they held a competition to write the scariest story, which led Mary to create *Frankenstein*. But it also inspired Byron to write the apocalyptic poem *Darkness*, in which the Sun is extinguished, and humanity gradually starves to death until only the poem's narrator remains. Over the next few years, several other poets picked up this "Last Man" theme; and ten years later, in 1826, Mary Shelley wrote *The Last Man*, commonly regarded as one of the earliest works of dystopian literature.*

The Last Man describes a plague that sweeps across the world in the late twenty-first century, and the struggles to survive of a group closely modelled on Shelley's real-life circle. Ultimately, all humanity succumbs to the disease and consequent social collapse, save only the titular Last Man—a proxy for Shelley herself. While poorly received at the time, it's now seen as prescient in several ways. For one thing, the pandemic's devastation is exacerbated by political bungling; the book also rebukes the idea that reason alone can ensure humanity's continued progress, suggesting nature has the upper hand. And unlike previous apocalypses, the end of humanity is not synonymous with the end of the world.

* It was not the first—*Le Dernier Homme* by Jean-Baptiste Cousin de Grainville was published in 1805.

This idea was to become more acceptable in the following decades, as it became clearer that things changing sometimes meant things *vanishing*. The realisation that fossils were the remains of disappeared, sometimes mighty, creatures proved that species can and do become extinct; the theory of natural selection provided a rationale for why that might happen. That suggested that people, too, might not be permanently sinecured as the rulers of the world. Even if humans were not actually expunged from the planet, civilisation might fall, and human life become nasty, brutish and short.

Dystopian literature picked up steam just as utopian literature did, but lagged by several decades; perhaps ideas have to be voiced as ideals first, before they can be interrogated more sceptically. Yevgeny Zamyatin's *We* (1921), Aldous Huxley's *Brave New World* (1931) and George Orwell's *Nineteen Eighty-Four* (1949) set the tone for much that followed. In later years, the line between dystopia and utopia became blurred, as futuristic fiction became more novelistic; "straight" utopias and dystopias ceded ground to more challenging hybrids.

Now straightforward dystopia seems to be back: from the likes of Cormac McCarthy's *The Road* (2006) to the return of Mad Max, gloomy dystopias abound. Perhaps that reflects our times. Shelley's book, which was out of print for more than a century, found a new audience (unsurprisingly) during the Covid-19 pandemic. Some readers also find it resonant with climate-change anxieties and solastalgia. Just as in "the year without a summer," the seasons are becoming blurred and confused; and we know that we can expect more to come: a permanent Mount Tambora.

One Damned Thing After Another

It's been said that history doesn't repeat, but it rhymes. You might expect certain broad patterns to recur, given the continuity of human nature and the turning of the seasons. But events like the eruption of Tambora are a reminder that patterns can be broken—and when they break, so do our worldviews.

If you were forced to pick a single event which defined the course of the twentieth century, the assassination in 1914 of Archduke Franz Ferdinand would be a good candidate. A shot fired in Sarajevo led to the outbreak of World War I; the smouldering resentment that followed ultimately ignited into World War II; the nuclear culmination of that conflict, and its uneasy settlement, led to the Cold War; and so on, down to the present day.

But the archduke's death need not have happened. His murder was the result of an absurd succession of accidents and mistakes, most of which could easily have been averted. He could have heeded warnings of the threat to his life; he wouldn't have been in an open-topped car if he hadn't wanted to be seen with his wife. He *might* have been killed earlier if the assassins had thrown their bomb more squarely; but then again, he might have taken the hint to avoid further danger. And above all, his driver might not have got lost and stalled the car in a blind alley—directly opposite the café where one of the assassins had gone to have lunch. The rest, as they say, is history.

Not quite. One factor that doesn't often make it into the textbooks is that Conrad von Hötzendorf, Franz Ferdinand's military chief of staff, was spoiling for a fight—not just because he was the bellicose product of a martial culture, but also because he was besotted with Virginia von Reininghaus, a married mother of six. Von Hötzendorf had somehow convinced himself that his Catholic paramour could be persuaded to divorce her husband if he pulled off a sufficiently impressive display of military prowess. In the event, he led the Austrian army to staggering losses and ultimately to swingeing defeat. But he did, in the end, win Virginia's hand in (re)marriage. So there was one happy ending to World War I, even if about a million Austrian soldiers had to die for it.

World War I presented a massive challenge to prevailing worldviews, just as the Lisbon earthquake had. In its aftermath, many grew sceptical about the idea of progress. In 1935, with a second world war looming, the Oxford historian Herbert Fisher wrote that there was "only one safe rule for the historian: that he should recognize in the development of human destinies the play of the contingent and the unforeseen. This is not a doctrine of cynicism and despair. The fact of progress is written plain and large on the page of history; but progress is not a law of nature." His eminent London peer, Arnold Toynbee, a pattern-seeker if ever there was one, responded waspishly. If you went along with Fisher, Toynbee wrote, history becomes no more than "one damned thing after another."

Toynbee meant to be dismissive. But what if history really *is* just one damned thing after another? And what would that mean for our hopes of reaching the best of all possible worlds?

Over the course of the nineteenth century, the natural sciences had gradually become accustomed to the idea that chance plays a major part in processes previously thought of as divine design. Geological cataclysms and climatic variations, evolutionary accidents and genetic mutations, chaos theory and quantum mechanics: all suggest that the world around us today was shaped by an uncountable number of actions and events, the vast majority of them having nothing whatsoever to do with human desires or decisions.

The social sciences have been slower to accept this principle, but it has gradually become more respectable to acknowledge the roles of happenstance, impulsiveness, accident, idiocy and lust. And with it has come the realisation that if things had gone differently yesterday—a different individual decision, a different social movement, a different political outcome—then today would have been different not just in terms of the consequences that flow directly from the alteration, but from an expanding web of knock-on effects.

If Franz Ferdinand hadn't been shot where and when he was (or if he had not been shot at all) the European powers might well have acted quite differently; World War I, if it happened, might have taken a very different form, and thus the world today would be quite different, too. And whoever lived in that world would not be *us*, transported sideways: they would be a different set of people with different genealogies, heritage, experiences and expectations. (Our fiction recognises this. If there's a golden rule for time travellers to the past, it's this: *Don't change anything*.)*

The study of such what-ifs is called counterfactual history, and it's been practised for almost as long as history itself. But counterfactuals came into their own in the nineteenth century under figures like the Prussian military strategist Carl von Clausewitz, who, writes literary scholar Catherine Gallagher, "explicitly established the counterfactual principle that to know the weight and significance of an event, the historian must . . . consider 'the full extent of everything that has happened, or might have happened.'" Von Clausewitz took this to great lengths: his analysis of Napoleon's options after his victory at the 1797 Battle of Valvasone ran to a "sixth-order hypothetical": what if A, *then* what if B, then what if C, what if D, and E, and finally, what if F?

There was no exploration of contributing factors off the battlefield, however: no questions about how an ill-timed attack might eventually, say, spiral into a battle for the balance of power in Europe. Questions like "What if your supreme commander falls in love with an unsuitable woman?" certainly didn't get a look in. But as time went on, those "irrelevant" questions did start to be asked. *If It Had Happened Otherwise*, a 1931 collection of essays, addressed such subjects as "If Byron Had Become King of Greece," "If the Emperor Frederick Had Not Had Cancer" and "If Lee Had Not Won the Battle of Gettysburg"—the last written by none other than Winston Churchill.

* Physicists have debated whether this really should be a rule, but since no one is very clear about whether time travel is even theoretically possible, the debate is more science-fictional than scientific anyway.

Mainstream historians' hearts did not warm to this experiment. One distinguished British historian dismissed counterfactuals as "idle parlour games"; another called them "unhistorical shit." Many professional historians remain sceptical. There's often little enough known about the past as it is. Why complicate matters further by throwing in things that definitely *didn't* happen?

Looking Sideways

"Why should we look to the past in order to prepare for the future? Because there is nowhere else to look," the legendary science writer and futurologist James Burke once said. But with due respect to my illustrious colleague, there is: we can look *sideways*. We can look to the choices that weren't made, the actions that weren't taken, the paths not trodden: the counterfactuals. We can ask the question we've all asked, at one time or another: "What if?"

We intuit that the future is a space of possibilities; we routinely make plans for what we'll do next: after this sentence, later this afternoon, tomorrow, next month, in the new year, when we retire, once we've got old. We dream of what we would do if we met Mr. or Ms. or Mx. Right, if we got our dream job or our dream house, if we won the lottery or went viral; we fret about what will happen if we *can't* find a partner, score a gig or make the rent. Such counterfactuals are the raw stuff of our daydreams, our career goals, our bucket lists and retirement plans.

By contrast, we tend to think of the past as having, well, passed. Its events seem fixed and immutable, and so they are. But the past is a space of possibilities, too—possibilities that are more symmetrical with those of the future than we may appreciate. We only get to live in one future; we only get to have one past. The paths not taken, whether they lie behind us or ahead of us, are equally imaginary in both cases, rearrangements of people and places in our minds. We only have imperfect memories and incomplete evidence from the past, just as we can only make imperfect predictions on the basis of incomplete expectations of the future.

The difference lies in our agency. We like to think we can, to some extent, control our futures. By contrast, we accept that our pasts are immutable: we can't change them no matter what we do. But our present was once the future; "today" was once "tomorrow." And it was just one of many *possible* tomorrows. Chance played a role in determining which one came to pass. But so did choice.

Counterfactual history reminds us that our ability to control the course of events is limited, liable to be thwarted at any minute by some unpredictable and unpreventable twist of fate. That's not to say we have *no* control; it's to say that we should recognise the limits of prediction and agency, and plan accordingly. If we conclude that the course is set not by some deep pattern of the world but by happenstance, we don't have to be fatalistic about the way things have turned out. We can be emboldened to think again about those possible worlds that were never realised—not because they fell short of providence, optimality or *Weltgeist*, but because of accident, conjunctures and curious juxtapositions. We can put together the pieces differently and imagine what might've happened instead—just as we construct scenes in our mind when imagining the what-ifs of our own lives. The way things were is not the way they had to be; the way things are is not the way they *have* to be. And thinking about the way things *weren't* can give us fresh perspectives on the way things *are*.

Consider, again, Austria before the assassination of Franz Ferdinand. "History from below" suggests that most people might have preferred to get on with their lives, but Europe, full of heavily militarised states, was a powder keg just waiting to blow. The archduke's assassination simply lit the fuse. If it hadn't been that day in Sarajevo, it would have been soon after. War was inevitable.

But as the historian Dan Snow has pointed out, you could make a similar case for the inevitability of the nuclear war which devastated the world in the twentieth century—except, of course, that no such war happened. "Had the USA and the Soviet Union gone to war in the second half of the twentieth century," writes Snow, "future historians, if there were any, would have wisely pointed out that war between these two superpowers, with multiple points of friction, armed as never before with giant arsenals with dodgy command and control mechanisms, and deeply antagonistic world views, was absolutely inevitable."

One explanation for why nuclear war between the United States and the Soviet Union remained a counterfactual comes from the theory of "mutual optimism." This holds that wars start when one or both sides believe they can gain more from victory in battle than through negotiation. Of course, one or both will be wrong. The nuclear standoff ramped up the stakes so greatly—to apocalyptic levels—that the likelihood of such gains appeared minuscule.

So no one has a rational reason to start a nuclear war, but the world has come perilously close a number of times. The bomb has nearly been launched for reasons including the misidentification of geese, the Moon and meteors as incoming missiles; a bear mistaken for a saboteur at an Air Force base in Duluth; errant Soviet submarines and American spy planes during the Cuban Missile Crisis; power outages and communication failures caused by solar flares and thunderstorms; faulty alarm systems, in one case caused by a defective forty-six-cent microchip; a "war game" simulation loaded into a live system; and a drunken order to bomb North Korea given by President Richard Nixon. Perhaps ours *is* the best of all possible worlds in that respect: it escaped immolation while its counterfactuals did not.

Of course, that's not purely down to chance. It survived because human action stopped an accident becoming an apocalypse.

Sometimes this involved clearing up a technical problem; but sometimes it was a matter of judgment. The most celebrated example of the latter is the case of Stanislav Petrov, the Russian lieutenant colonel who in September 1983 saw a US missile launch on the Soviet detection system he was monitoring, followed by four more.

It was a time of high tension between the superpowers. A series of skirmishes and misadventures—accidents—meant trust between the US and Soviet leadership was at rock bottom. Provocative military manoeuvres were taking place and nuclear weapons had been deployed in new frontline locations across Europe. The Soviet Union had adopted a "launch upon warning" doctrine: once a launch had been detected, the computer stood ready to fire in immediate retaliation. "The main computer wouldn't ask me," said Petrov later. "It was specially constructed in such a way that no one could affect the system's operations." Once given the START command it would simply fire away.

Petrov, however, thought it unlikely that the US had decided to attack with just five of its hundreds of missiles, and felt it more likely that the detection system was malfunctioning. So he decided to disobey his explicit orders and didn't give the START command. A tense few minutes later it became apparent he had been right to do so: no missiles fell. Under these highest of stakes, he had evaluated the possible worlds and chosen the most benevolent possibility: the most favourable construction of events. His optimism had saved the world. Accidents happen. But human agency has the power—sometimes—to prevent them.

Killing Hitler

It's comforting to think that God, or *Weltgeist*, or Enlightenment values are propelling us towards a glorious future. But let's for the sake of argument assume there's no interventionist God, no destiny and no plan. This was the worldview championed by Jean-Paul Sartre, everyone's favourite chain-smoking French philosopher.

"We are left alone, without excuse," wrote Sartre in 1946. "That is what I mean when I say that man is condemned to be free. Condemned, because he did not create himself, yet is nevertheless at liberty, and from the moment that he is thrown into this world he is responsible for everything he does." This is a sentiment that Voltaire might have agreed with in one of his darker post-Lisbon moments. Many found it too bleak to accept. It stripped humanity of dignity, Sartre's critics argued: What difference was there in his philosophy between a human being and an inanimate stone, or an insensible vegetable? Simple, replied Sartre: "Man is, before all else, something which propels itself towards a future and is aware that it is doing so." In a world without God, we are what we do, and the future is what we make it.

But Sartre argued that by the same token, we can't pretend our disappointments are due to anyone but ourselves. If we didn't make it as a pro football player, write that novel or secure a long and happy marriage, it's on us—not God, genetics or the universe. If we want to be our best possible selves, if we want to live in the best of all possible worlds, it's incumbent on us to make it so. "In the light of all this, what people reproach us with is not, after all, our pessimism, but the sternness of our optimism," he concluded.

Stern, but effective. Sartre was writing in the immediate aftermath of World War II. The French reaction to the conflict, and the evils it had spawned, could have been despair, their faith in God and man shattered just as their Catholic predecessors' faith had been shattered by the Lisbon earthquake two centuries earlier. Sartre—who enjoyed a greater public profile in his day than most philosophers, past or present—has been credited with providing an alternative: a worldview that confronted the reality of the human condition without resorting to faith.

There's plenty of scope to disagree with that worldview. Few would agree that we are *entirely* free to act, nor that we are *fully* responsible for our circumstances. Our actions are shaped by our lives and our environments, and by our social and economic circumstances. Some things happen to us that we don't have any control over. But it does suggest that it is possible to be optimistic without belief in God, or for that matter without putting our faith in a greater power of any sort. We might find it liberating to feel that our fate is in our own hands, not the incomprehensible plan of an aloof and unaccountable deity. But it's also a responsibility, and we need to work out what to do with it. Counterfactuals can help.

By their nature, counterfactuals become more popular when we are struggling to understand the world. While there are a fair number of counterfactuals dealing with what might have happened if Archduke Franz Ferdinand had *not* been shot in 1914, the form was then in its infancy and most were created long after the event. But by the time World War II came round, alternate histories were on their way to becoming a well-established genre, and the questions they raised were more timely.

"Would you go back in time and kill Hitler as a baby?" is probably the most widely discussed of all counterfactual thought experiments. Would such an act be cold-blooded child murder, since the infant Adolf has, as yet, done no wrong? Could we not instead change his formative years so as to make the grown man less hate-filled? The Great Man theory says Hitler is key to how World War II turned out; history from below says social pressures would still turn Germany to Nazidom. So: would you pull the trigger? (In 2015, 42 per cent of respondents to a *New York Times* magazine poll said they would.)

There are many fictional alternate histories imagining a world where the Axis powers won the war, the ur-text being Philip K. Dick's 1962 novel, *The Man in the High Castle*, set in a United

States occupied and divided between Imperial Japan and Nazi Germany.* Numerous thrillers followed, ranging from Philip Roth's *The Plot Against America* to Newt Gingrich's novel *1945*—published the same year as his conservative manifesto *To Renew America*. "In this world, the Nazis won" has been a stock premise of books, TV, films and games for decades, and there's no sign of it stopping.

The social function of these what-if stories has changed over time. Initially, they were triumphalist narratives "explaining" how national defeats were actually victories in the bigger scheme of things: very Leibnizian. Later, they became more contemplative attempts to redress the ills of the past. Roth's book, for example, acknowledges that antisemitism was rife across the western world, not just Germany, during the early twentieth century. Gavriel Rosenfeld, editor of *The Counterfactual History Review*, argues "nightmare" scenarios, such as an occupied America, reassure us that the correct decisions were made back then; fantasy scenarios, that things could have gone better.

The pattern repeats in the nineties, when serious historians finally started to warm to counterfactuals. Many restricted themselves to military questions, but some occasionally ranged further afield, as in 1997's *Virtual History: Alternatives and Counterfactuals*, edited by the historian Niall Ferguson. In his introduction, Ferguson argued that this approach neatly illustrated how actual history is a chaotic sequence of events, capable of multiple interpretations. Not everyone agreed: in a repeat of the Fisher–Toynbee spat, the distinguished historian Richard Evans beefed with what he saw as the counterfactualists' tendency to "rewrite history according to their present-day political purposes and prejudices."

* There's a convention that works intended as fictional entertainment are known as "alternate" histories, while those intended as intellectual exercises are "counterfactual." In practice, they can be hard to differentiate.

Why did counterfactuals and alternates catch on in the nineties? As Evans put it, ideologies like fascism, communism, socialism and Marxism had ceased to be "*isms* and become *wasms*" instead. Catherine Gallagher suggests it was the sudden demise of "the world-order-as-we-knew-it," following the collapse of the Soviet Union in 1989. "The grip of historical necessity had loosened and released us from narrative coherence," Gallagher writes. "Historians, policy-makers, international studies and political science scholars, not to mention world leaders, were all caught seriously unprepared, unable to explain either what happened or why they had not seen it coming. The old scenarios were scrapped as planners scrambled to find the organizing principles of the new reality."

If counterfactuals rise in popularity after a sharp change in worldview, are we due another increase now? That does, in fact, seem to be happening. Counterfactuals are everywhere—as we saw in the last chapter—and focused on the increasingly recent past.

In the "Remedial Chaos Theory" episode of the sitcom *Community*, which aired in 2011, one character rolls a die to determine which of the regulars needs to go for pizza. "You are now creating six different timelines," the show's resident science-fiction enthusiast, Abed, warns, and we're then shown what happens in each of them. In the one where Abed's friend Troy goes to get the pizza, the result is a succession of *Candide*-style calamities, culminating in one character getting shot and the apartment catching fire. This, Abed says, is "the darkest timeline," a reference the show continued to call back to for the rest of its run.

"The darkest timeline" went on to become a rolling meme among *Community* fans and various online communities—but broke out when Donald Trump was elected in 2016. References to the darkest timeline began to sprout wherever alarmed liberals gathered, sometimes modified to "the dumbest timeline," or other snowclones of the phrase.

Some went further. In his time-travel novel *Agency*, published in 2020, William Gibson wrote of an alternative 2017 in which Hillary Clinton won the election, and the British people voted to remain in the European Union. Gibson had been working on a near-future novel, but, he says: "I woke up after the presidential election and realised the world I had set the book in no longer existed." Feeling that reality had shifted from where it "ought" to have been, his reworked novel included history as he felt it *should* have been.*

Wish fulfilment? Perhaps. But Gibson was not alone in producing a counterfactual so close to home: Curtis Sittenfeld's novel *Rodham* imagines what would have happened if Hillary had pursued her own career rather than supporting Bill's, interweaving through the history of American politics. There are UK counterfactuals involving living politicians, too: an entire series, in fact, from Biteback Publishing, with titles like *The Prime Ministers Who Never Were*. Another, published in 2011, is boldly titled *Prime Minister Boris: And Other Things That Never Happened*. But it did happen, in 2019. Come close enough to the present, and the counterfactual can merge with the actual.

History may have its patterns, but patterns can be broken. Some accidents of history are purely objective events: a volcano explodes. The interpretation of others depends on where you stand in relation to them: the morality and significance of killing Hitler changes with time and context. Still others, like President Hillary, fall into the category of wishful thinking—not to be confused with possible, or plausible, but potentially instructive all the same. Progress may or may not march on; but if it does, it will occasionally break stride.

And that, in fact, is why optimism—optimism without—necessarily exists. If the world only ever proceeded in a smooth

* In fact, Gibson's 2017 is an alternate reality *within* the novel, too: its "real" timeline is one established by its sibling novel, *The Peripheral*. This kind of matryoshka-doll tricksiness is a hallmark of the genre: *The Garden of Forking Paths* and *The Man in the High Castle* both toy with multiple alternatives.

trajectory, or a regular cycle, it would be entirely predictable, and we'd have no need of expectations, positive or negative. We'd have no need to be optimistic, or reason to be pessimistic. But as it is, accidents *do* occur; and when they do, we can only hope that our agency is sufficient to overcome them. That in turn requires us to predict the future with some degree of accuracy. But can we?

6

The Elusive Future

Can we see into the future—and can we control it?

Oedipus, prince of Thebes, should have led a charmed life; instead, he endured a cursed one. His father, King Laius, had been told by the famous Oracle at Delphi that he would one day be slain by his own son, by way of comeback for the mortal sins he'd committed in his own youth. Heeding the warning, he tried to avoid getting his wife pregnant, but one night wine was taken, and events took their course. When the infant Oedipus arrived nine months later, Laius ordered him to be taken from the palace and abandoned.

The servant charged with that duty, however, didn't have the heart to go through with it: instead, the child was delivered to the royal house of Corinth, where he was raised as a prince. Once he reached adulthood, Oedipus, too, consulted the Oracle, who told him he would kill his father and marry his mother. Horrified, and determined to escape this fate but unaware of his true parentage, Oedipus fled, only to find his way blocked by a chariot. In the ensuing quarrel, he killed both the chariot's driver and its passenger.

Wandering in self-imposed exile, Oedipus encountered the monstrous Sphinx near Thebes, his birthplace. The Sphinx devoured travellers who could not answer her riddle: "What walks on four feet in the morning, two in the afternoon, and three at night?" Oedipus correctly replied that it was man: crawling as a baby, walking as an adult and hobbling with a stick in old age. The Sphinx was defeated, and Thebes's grateful populace awarded Oedipus its throne and its queen's hand in marriage

You can guess the rest if you don't know it already. The passenger in the chariot was Oedipus's father, Laius; the Theban queen was his widowed mother, Jocasta. When the truth emerges, Jocasta kills herself, while Oedipus puts out his own eyes.

The Oedipus myth is about our ability to foresee and understand the world. The definitive version is Sophocles's play *Oedipus Rex*, first performed around 429 BC and declared by Aristotle in his authoritative *Poetics* to be the very greatest of Greek tragedies. One of the key components of a tragedy, explains Aristotle, is the protagonist's *hamartia*—the character trait, usually a "tragic flaw," that propels them to their fate. Various mundane candidates have been put forward for Oedipus's *hamartia*, including his temper and his impatience. But my preferred answer is that it's his unwillingness to accept his destiny. In the world of Greek tragedy, mortals are powerless to escape their fates, particularly if those fates are decreed by the gods. The Oracle can see the future; the Sphinx's riddle hints that it's fixed; but Oedipus and his family continue to act as though they can direct the courses of their lives, to tragic effect. The future is immutable, and there's no point trying to avoid or subvert our divinely ordained fate.

In such a world, there's no point trying to make things better, or to improve your life, or in fact in trying to achieve *anything*, given that whatever is fated to happen will somehow happen anyway. Aristotle called this the "Lazy Argument": Why bother trying? You might as well do whatever you want. Under this reading, Oedipus's *hamartia* is his belief that he can evade his fate—an expectation the Greeks would have viewed as hopelessly unrealistic. A positive illusion. Remember that optimism tilts us towards action, pessimism towards fatalism. Oedipus is an optimist in a fatalistic world, and that is what dooms him.

This is quite alien to the modern western worldview. "The Judeo-Christian view of the world must contain hope—of resurrection, of redemption, of forgiveness . . . it ends well. Tragedy ends badly," explained the classicist George Steiner. "The Greeks said the best thing is not to be born, after that is to die very young, and the worst is to live long . . . This midnight view that we are somehow guests on an Earth that is hostile, or enemies of the gods, or objects of vengeance . . . that's a pretty bleak doctrine."

But fatalism was to make a comeback. Since the late Middle Ages, philosophers had compared the universe to a clock set in motion by some "first cause" (which is to say, God); the cosmos had run as surely and mechanically as clockwork ever since. That idea gained momentum with the discovery of universal principles—Newton's laws of motion, among others—which suggested that the world could, indeed, run like clockwork with no assistance from its creator. Leibniz was a strong proponent of this "deterministic" view; Newton himself was not, which was another cause of the discord between the two men.

This resurgence also shows up in literature. As it happens, Voltaire's first big success was an adaptation of *Oedipus*—and his first big scandal arose from a gender-flipped version of the same story that hinted at an incestuous relationship between the French regent and his daughter. The play made him the doyen of the Parisian stage; the poem got him imprisoned in the Bastille. And to some extent, Oedipus's theme is recapitulated in *Candide*: an innocent abroad endures great torment only to find that the world will, in the end, be however it will be—regardless of your efforts.

By the late eighteenth century, many thinkers were satisfied that our universe is indeed deterministic. One of them was the philosopher Arthur Schopenhauer, who argued, following Leibniz's lead, that if there was "sufficient reason" for something to happen, then it *must* happen. If there is a cause, there must be an effect. And that implies that we have no ability to make decisions or choices of our own. We might think otherwise, but this is an illusion, just as it was for Oedipus. We have to accept whatever the world has in store for us—and for Schopenhauer, that means suffering. You might argue that there is joy, too, but as he famously argued in his 1851 essay "On the Suffering of the World": "A quick test of the assertion that enjoyment outweighs pain in this world, or that they are at any rate balanced, would be to compare the feelings of an animal engaged in eating another with those of the animal being eaten." Pleasure, meanwhile, is merely the cessation of pain—you eat to prevent yourself being hungry—and so pain has primacy.

So existence inevitably means suffering; it would be better never to have been born. But since we do not all take our own lives immediately, our lives must be bearable, but only *just*—which suggests that this is actually the *worst* of all possible worlds. Now *that's* pessimism.

The Block Universe

The physical laws which supported the clockwork universe have long since been superseded. Physics today is rooted in relativity and quantum mechanics. The former bends time and space; the latter reintroduces uncertainty and randomness. But these pillars of modern cosmology don't suggest the universe has any interest in our struggles or successes either: quite the opposite. The picture of the cosmos that emerges from them affirms that the universe is not even in motion, like clockwork, but entirely static—like a frozen block. Time doesn't "flow" in this picture; we don't move from the present to future. We might *feel* as though it does, and we do, but that's just another illusion. We can go through the motions, but ultimately everything we've done or will do is already fixed.

This "block universe" is pretty hard to relate to human experience, but it has devoted followers. When Albert Einstein's great friend Michael Besso died in 1955, he wrote to Besso's family: "Now he has departed this strange world a little ahead of me. That signifies nothing. For us believing physicists, the distinction between past, present and future is only a stubbornly persistent illusion." It is also purely theoretical: we currently have no way of establishing if that is really how the universe is arranged. While many physicists do believe it, they either shrug at it or begrudge it. Some believe it *must* be incorrect.

The point here isn't really what the block universe says about human existence. It's that where Leibniz could fall back on God's divine plan to take the edge off determinism, we have only this discomfiting picture of cosmic indifference. That has helped both philosophical and popular pessimism to make a comeback. The philosopher David Benatar, for example, suggests that it refutes any suggestion that life is meaningful; as Schopenhauer argued, existence inevitably means suffering and therefore having children is fundamentally a mistake. For the title of his 2006 book on the subject, Benatar returned to the Greek sentiment: *Better Never to Have Been: The Harm of Coming into Existence*.*

You don't need to be either a "believing physicist" or a pessimistic philosopher to hold this view: you just need to feel that existence lacks purpose—a feeling which may not be that hard to arrive at when God is dead and the world is on fire. Anti-natalism can't be said to be a mainstream idea, although there are niche communities dedicated to it, and it had a viral moment in 2019, when a twenty-seven-year-old Indian man called Raphael Samuel told the BBC that he was going to sue his parents for having conceived

* Benatar's specific wording comes from Sophocles's sequel, *Oedipus at Colonus*. As the now-blind king awaits his death, having been cursed yet again, the chorus sings: "Never to have been born is best / But if we must see the light, the next best / Is quickly returning whence we came."

him without his consent. But it is of a piece with Les Knight's Voluntary Human Extinction Movement, and with the concerns we've examined about having children in a time of environmental degradation. If optimism urges us to action, and having children is our most optimistic act, then choosing *not* to have children because doom inevitably awaits is a return to the midnight view.

I think we can do better than that. We are organisms made of flesh, blood and bone, whose minds have been shaped by millions of years of evolution to survive an environment that was often doing its best to kill us. To put it as Leibniz might have done, we have only our creaturely perspective to go on; our ability to grasp the cosmic perspective, and draw lessons from it, is at best acutely limited. Our ability to anticipate and take control of our futures, whether illusory or not, has seen us rise to challenges that would have instantly defeated any other animal.* The real question is not whether we can *actually* change the future. We may never know the answer to that. The question is whether we *believe* we can.

Some of us return to the old ways to restore that belief. Some years ago, I reported for the BBC on the potential for new religious movements: neopagans calling on old gods to defend the land; technological cults dedicated to the promotion (or prevention) of godlike machines; secular congregations preaching godless unity and celebration. Many of these people didn't believe in any supernatural organising principle, but did believe in the power of ritual and community—attempting, to use Zimbardo's terminology, to deliberately shift their time perspectives from present-fatalist to a mix of past-positive and future-oriented. When you feel helpless to change your fate, call upon a deity for assistance. Or invent one.

* Some neuroscientists argue that our sense of agency is illusory for a different reason: because what we perceive as decisions and actions made of our own free will are actually the result of subconscious, automatic processes. But this suggestion is even more counterintuitive, and contentious, than the block universe.

I think, though, that this is another way to cultivate positive expectations—to cultivate optimism. It provides psychological comfort in the face of a world that seems to be spinning out of control. After all, we don't know what the future holds. We don't know what effect our choices and decisions will have on it. Worse still, we don't know what the choices and decisions made by *other* people—bosses, partners, family, friends, even total randoms—will have on our lives. We don't know if our jobs might be taken by a computer, the world will ignite or civilisation collapse.

We can, of course, just hope that all these things will sort themselves out. But that's *blind* optimism—the blindness of Oedipus, trapped by fate. Or of Pangloss: simply accepting that all must be for the best in the best of all possible worlds. Better to at least *try* to see the possible worlds of the future, and then to try to control them.

The Sea Battle

Let's go back to first principles. There are no facts in the future. If statements about the future really *were* facts, if they were about events that were definitely going to happen, that would imply the future had already been determined. That might not seem particularly problematic when it comes to, say, the Sun coming up tomorrow. We want and expect that to happen, and we don't expect to have any sway over it. But in other cases, it would be more problematic.

Aristotle, in his treatise on logic, *On Interpretation*, gave the more dramatic example of a sea battle. If we can say *for sure* that a sea battle is going to happen tomorrow, then nothing we do today can affect it. Nor could anything we did yesterday, or the day before that, and so on back to the end of time. The sea battle will always have been destined to happen: no one could ever have stopped it. Conversely, if we could say for sure that it *wasn't* going to happen, no one would ever have been able to start it, no matter what provocation they offered. In the terminology of modal logic, it would be either necessary or impossible. Human agency would be meaningless.

You don't need to be into maritime warfare to see a problem here. We'd like to think we can prevent wars breaking out. We'd like to think we *have* prevented wars breaking out, as Stanislav Petrov did when he ignored his orders in 1983. We want to believe in human agency, not in some cosmic plan of which we know nothing. That would be, well, tragic. "There is nothing true in Oedipus' life at the moment of his tragic fall that was not also true every single moment of every prior day," writes the scholar David Kornhaber.

Rather than accept the fatalist implication of classical logic, Aristotle proposed a get-out clause: the sea battle would be neither necessary nor impossible, but *contingent*: neither true *nor* false. This is perfectly in line with intuition. What we know about the world, we know through the evidence of our senses. It's easy to establish whether the statement "It is raining" is true or false: you can look out of a window, or put an exploratory hand out of it. We can say "It will rain tomorrow," but we know that might turn out to be either true or false. On the other hand, to say "It *is* raining tomorrow" is nonsensical: we can't put our hand out of tomorrow's window. The most accurate statement we can make is "It *might* rain tomorrow," which, if you think about it, doesn't really say anything at all.

Our choice, on the face of it, is between knowing nothing and changing nothing—between ignorance and fatalism. Thinkers have been trying for millennia to find a third way. Some have built logical systems which include a third alternative to "true" or "false." These have their uses when it comes to making computers deal with the "fuzzy logic" of the real world. But as anyone who's ever asked for a date or a loan knows, "maybe" is a frustrating answer.

Others have suggested that various combinations of the past, present or future don't actually exist. *Presentists* say that this moment, right now, the moment in which your eyes are scanning these words, is all that there is. Now that moment is in the past, just as that sentence is no longer in your view. And now you've arrived at *these* words, in the future that didn't exist just a moment ago. *Eternalists,* by contrast, say that everything already exists, just as all the words in this book already exist, even if you only remember the ones on the pages before this one, and can only guess at what's on the pages you haven't arrived at yet.*

But one of the most widely favoured approaches was first suggested by the fourteenth-century English theologian William of Ockham. He hit on the idea that while there might be many *possible* futures, there was only one that would *actually* come to pass, and only God knew which one that was. That might sound familiar. Leibniz tackled the problem of future contingents in his own time by positing that there was one true future, selected by God out of all possible futures—some decades before he suggested that optimism meant God had selected one world out of myriad possibilities to create.

This view has its own problems—principally that if only one of all these futures ever comes to pass, and only God knows which one, in what sense are the other futures really "possible"? But it is how we tend to think about the future today: as a space of possibilities yet to be realised. We have that functionality built into our brains, the result of millions of years of evolution: not only can we conceive of numerous futures for ourselves and for the world, but we can evaluate and choose between those conceptions.

* Trust me, they're great—much better than this one. Keep going.

And we've built tools that augment our natural abilities to do that: instruments, mathematics, science and computers allow us to make precise predictions of what's coming next.

The Forecast Factory

Will it rain next week? The weather depends on numerous meteorological factors—air pressure, wind direction, humidity, temperature and so on—whose interactions are all but impossible for an unassisted human being to evaluate. You might make a guess based on recent weather patterns, your location and the time of year, but you'd hardly be surprised to find you'd got it wrong. But just as we've used science and technology to further the reach of our senses in space, we've also used them to increase our reach in time—to figure out which of the possible futures is the most likely to come about.

In 1922, the British mathematician Lewis Fry Richardson published a book called *Weather Prediction by Numerical Process*, in which he proposed dividing up the world's surface into "cells," running equations describing atmospheric phenomena for each cell, and then stitching these together to get an idea of the weather that would result. This, he suggested, would happen in a vast "forecast factory":

> Imagine a large hall like a theatre, except that the circles and galleries go right round through the space usually occupied by the stage. The walls of this chamber are painted to form a map of the globe. The ceiling represents the north polar regions, England is in the gallery, the tropics in the upper circle, Australia on the dress circle and the Antarctic in the pit. A myriad computers are at work upon the weather of the part of the map where each sits, but each computer attends only to one equation or part of an equation.

By "computers" Richardson meant not machines, but human mathematicians—a cool 64,000 of them.* This massive team would be directed by a conductor-like figure using a system of lights, communicate with each other via mail sent through pneumatic tubes, and send out their forecasts by radio. Needless to say, that isn't how it's done today. But the basic principles of his system were correct.

Predicting tomorrow's weather—or the behaviour of any complicated system, from the movements of the planets to the growth patterns of plants—requires us to first *model* it, perhaps by finding an equation that seems to generate the same behaviour, perhaps by building a simplified version of it to find out how its key components interact, perhaps by building a simulacrum of it, whether that is a physical miniature or a "digital twin." Such models can be used retrospectively to check whether they match past results (hindcasting) or prospectively to see if they can predict the future (forecasting).

Today, human "computers" with slide rules have long since been supplanted by supercomputers that model the behaviour of the atmosphere in cells, as Richardson suggested, using equations that have been elaborated, but are still fundamentally the ones he had in mind. These models are populated with vast quantities of data, from sensors mounted on towers across the planet, on balloons floating in the air, or on flotillas of orbiting satellites. Thanks to continuing progress in both theory and practice, a five-day forecast today is as accurate as a one-day forecast was in 1980; we've also become better at providing timely warnings of dramatic events like torrential rain and hurricane landfall. "It will rain tomorrow" now *feels* more like a statement of fact than a prediction.

* This use of the term "computer" has been more widely understood in recent years, with particular reference to the people who performed the calculations needed for the Apollo space programme—first by hand and later by programming early electronic computers. In meteorology, as in astrodynamics, many of these computers were women whose contributions went largely unacknowledged.

The value of this hard-won matter-of-factness is enormous. Routine forecasts save thousands of lives, and early warnings of extreme events can save tens of thousands, while reducing uncertainty in disciplines ranging from agriculture to warfare generates economic benefits worth an estimated $162 billion each year. There's a limit to how good these forecasts can get, though, thanks to the notorious "butterfly effect," which was discovered by MIT meteorologist Edward Lorenz in 1963, but only really noticed when he gave a talk in December 1972 entitled "Predictability: Does the Flap of a Butterfly's Wings in Brazil Set Off a Tornado in Texas?"* Indeed it could, he explained, or equally it could prevent one from starting up. What Lorenz had discovered was that a tiny change in the starting conditions could lead to massive differences down the line, making them fundamentally unpredictable: *chaotic*.

Weather forecasters today get around this problem by generating *ensembles* of forecasts—running their simulations repeatedly with slightly different inputs and comparing the outputs to see if any consistency emerges. Or to put it another way, they generate a huge number of possible worlds, then check to see if they look similar. If most of them do, there's a reasonable chance that the actual weather will follow suit. Not so much seeking the best of all possible worlds as the most *common* of all possible worlds. But it's in the nature of a chaotic system that it will occasionally go far *off-piste*. That's why the forecast for tomorrow is still, occasionally, wildly incorrect. And although we can predict with some confidence that it will rain tomorrow, or even next week, we *can't* say if it will rain a month from now—to the dismay of wedding planners everywhere.

What about the climate? The global climate is an enormously complex system comprising the planet's air, water, ice, rock and life.

* Originally, Lorenz had described the instigating factor as the flapping of a gull's wing, but his colleagues suggested a more poetic metaphor would meet with greater success. They were evidently correct.

Each of these is a complex system in itself, and their behaviour and interactions are complex (and sometimes chaotic), too. In short: it's complicated. But as with weather forecasts, the key is ensemble forecasting: trying the model repeatedly with slightly different starting conditions gives an idea of the potential range of variation in the climate in years to come. Because climate is a large-scale, long-term phenomenon, and the initial conditions don't vary all that much—the Earth stays in much the same orbit, capturing much the same amount of energy from the Sun—it is generally fairly stable for fairly long periods of time, by human standards.

Sometimes, however, the climate does change markedly, as it did during the year without a summer; it can also be more protracted and regional, as it was during the "Little Ice Age" that afflicted Europe between the sixteenth and nineteenth centuries. Ours is one of those times, global rather than local, and enduring rather than transitory, and of course the culprit is our activity. We are the new variable in the climate models, and while the consequences are impossible to predict precisely, we do have some idea. As my friend and colleague Michael Brooks puts it, you may not be able to predict the route a ball will take through a pinball machine, but you can predict that the average score will change if you tilt the whole machine. If you pump greenhouse gases into the atmosphere, you can be confident that the average temperature will go up.

In fact, you can build a more sophisticated model—much more sophisticated—which won't produce a prediction, but will produce likely *scenarios* under different assumptions about the level of emissions, and how the various climate systems will respond. Just as with the weather, the most likely scenarios become the basis for planning. And we need that planning, because when it comes to the climate, *any* significant change from the norm will be problematic for agriculture, societies and industries built according to the expectations of a different time. But the further we deviate from historic norms, the more likely it is the system will respond in unpredictable ways.

One thing that's almost certain is that the gains we've made in weather forecasting will come to a halt, and perhaps go into reverse. At the moment, we can make decent forecasts out to about ten days before chaos kicks in. As the globe grows warmer, that horizon will decrease: in 2021, researchers at Stanford University estimated that the time horizon for our ability to predict rain would shorten by about a day for each three-degree rise in temperature. That level of heating is at the upper end of what's likely, but our ability to make forecasts is likely to be compromised across the board: we may already be seeing this as "freak" weather events become more evident. As yet, there's not enough data to tell if they're happening more often than the models suggest.

There are ominous signs, though. During the Antarctic winter of 2023, the sea ice that had so implacably trapped Shackleton's ship a century earlier simply failed to reappear. The extent of the ice cover was what would be predicted to occur just once in 2.7 million years, according to the standard models based on historical experience. But this is really a measure of the models' inadequacies in our changing times (and climes). Left to its own devices, the Antarctic sea ice might only drop to that level once every 2.7 million years—but it hasn't been left to its own devices. The chances of such an event have been boosted enormously by our activities.

Another possibility is that we will push a critical element of the climate past a point of no return: that the Amazon rainforest will experience a mass die-off, the Antarctic ice sheet will undergo catastrophic melting, or the system of currents that carries warm water from the tropics to the north Atlantic will collapse. Any of these could prompt an abrupt "phase transition" in the global climate—a dramatic change from one climatic regime to another. No such transition would be remotely positive for human civilisation.

One worst-case scenario is that the entire Earth's climate could become genuinely chaotic: one year might be riven by extreme swings from hot to cold; the next might be uniformly calm; a third would remain unfeasibly hot. Adapting to such a world would be extremely challenging, to put it mildly: you have to be ready for anything, anytime, anywhere. And you would have little idea what was coming next.*

I should stress that this is nowhere near being one of the consensus scenarios of climate modellers; the team that came up with it were seeking to find out if it was even possible, and if so, what might bring it about. As I write this, it is still being reviewed. Disquietingly, though, they concluded that it *was* possible, and that the tipping points were poorly understood but potentially closer at hand than might be expected. Pass one, and that would again be grounds for fatalism—not because everything is fixed and determined already, but because nothing is. Human agency is worthless in a world where *everything* is an accident. It's impossible to be an optimist in a world where you have no ability to change what's coming next. It's also impossible to be an optimist in a world where you have no ability to predict what's coming next.

The Population Bomb

For the moment, we *can* see what's coming next, some of the time, and we can change it, some of the time. That ability shapes our twenty-first-century optimism: we are neither as impotent nor as blind as our ancestors. We can't stop earthquakes, but we can often anticipate them. But there are limits to our methodological advances, especially when we move away from the physical and into the social. We can learn something about those—and by contrast, about the power of human agency—by looking at a previous occasion on which we faced down environmental doom.

* This is essentially the issue faced by the aliens in Cixin Liu's science-fiction blockbuster *The Three-Body Problem*: their planet's three suns create a chaotic system.

"The battle to feed all of humanity is over," declared the 1968 book *The Population Bomb* uncompromisingly—and not because humanity had won. "In the 1970s and 1980s hundreds of millions of people will starve to death in spite of any crash programs embarked upon now. At this late date, nothing can prevent a substantial increase in the world death rate."

The Population Bomb was written by American husband-and-wife biologists Paul and Anne Ehrlich, although only Paul was credited, at the publisher's request. In its first chapter, "The Problem," he describes the epiphany he experienced while crawling through a New Delhi slum in a decrepit cab "one stinking hot night," its streets alive with people. "People eating, people washing, people sleeping. People visiting, arguing, and screaming. People thrusting their hands through the taxi window, begging. People defecating and urinating. People clinging to buses. People herding animals. People, people, people, people."

At the time, Delhi had just under 3 million inhabitants; Paris had around 8 million. And while India had indeed suffered repeated famines over the previous two centuries, there had only been one in the twentieth: the Bengal famine of 1943, which killed up to 3 million people and is widely considered to have resulted from the redirection of rice to the British war effort in Southeast Asia. Worldwide, such disasters were often triggered by crop failures, but exacerbated and prolonged by political strife or military conflict. Population wasn't in itself considered a major factor.

But the Ehrlichs concluded that so many people could not be supported for much longer. Americans had a responsibility to help "our less fortunate fellows on Spaceship Earth" out of their "plight"; without rapid, global action, hundreds of millions would starve in the coming decades. That action included radical measures to dissuade population growth, including tax penalties for large families and withdrawal of aid from countries "unable" to become self-subsistent. (President Lyndon Johnson had threatened India with just such a withdrawal in 1966, encouraged by hawks who suggested that the subcontinent's growing population represented a threat to trade and national security.)

More startlingly, the Ehrlichs suggested adding chemicals which cause temporary sterility to drinking water and basic foodstuffs. Even the most ardent opponent of famine might have balked at that, given the disgraceful history of "population control" measures being targeted at those deemed mentally, physically or socially "unfit." The history didn't put the Ehrlichs off—the book even includes a fleeting reference to the potential necessity of coercive sterilisation if voluntary measures prove insufficient—but the science did. "Those of you who are appalled at such a suggestion can rest easy. The option isn't even open to us . . . thanks to the criminal inadequacy of biomedical research in this area," they wrote. "If the choice now is either such additives or catastrophe, we shall have catastrophe."

Initially, *The Population Bomb* attracted little attention. But Paul Ehrlich was an energetic and charismatic pitchman for his ideas, which eventually broke through to the general public. The book became a runaway bestseller, and Ehrlich, now something of a celebrity, was invited to appear on TV talk shows and news broadcasts as a prophet of doom. He appeared on *The Tonight Show*, for example, no fewer than eighteen times. Millions heard, read and accepted his version of events, as well as his prescriptions for avoiding disaster—including Les Knight, who was so convinced that he got his vasectomy and started calling for the human species to extinguish itself.

The Ehrlichs' prescriptions were taken up in some places. "Some population-control programs pressured women to use only certain officially mandated contraceptives. In Egypt, Tunisia, Pakistan, South Korea and Taiwan, health workers' salaries were, in a system that invited abuse, dictated by the number of IUDs they inserted into women," Charles Mann wrote in *Smithsonian Magazine* in 2018. "In the Philippines, birth-control pills were literally pitched out of helicopters hovering over remote villages. Millions of people were sterilized, often coercively, sometimes illegally, frequently in unsafe conditions, in Mexico, Bolivia, Peru, Indonesia and Bangladesh." Nor were such policies restricted to poor countries. Sweden forcibly sterilised around 20,000 people between 1906 and 1975; in 2024, 143 Greenlandic women sued the Danish government for fitting them with contraceptive coils without consent in the sixties, some when they were still young teenagers.

The world's two most populous countries also embarked on mass birth control. The Indian government undertook a huge vasectomy drive in 1976, sterilising millions of men under circumstances that often smacked of coercion: fare-dodgers on trains could get the snip rather than paying an unaffordable fine, for example. The programme subsequently turned to women, despite the greater complexity and risks of female sterilisation, because it was assumed women would put up less resistance. Mass sterilisation persists in India to this day—as do concerns about safety and consent.

China took a different approach, introducing a one-child policy in 1979 which it claims prevented around 400 million births. Again, there were allegations of coercion and of unfairness, since many groups were either exempt from the programme or able to buy their way out. One such group was rural families whose first child was a girl; boys were more desirable, would carry on the family lineage and were deemed more capable of working the fields. Both China and India have ended up with marked gender imbalances as a result of such cultural preferences.

In short, many population-control measures were carried out under circumstances we'd today consider to be in violation of human rights. At the very least, a vast number of people were deprived of the choice to determine their own lives and destinies. Whether there was ever a pressing need to control populations is debated: it's been claimed, for example, that China's one-child policy was a politically acceptable and seemingly scientific alternative to admitting that the country's sluggish growth might have other causes.

You already know what happened next. The Ehrlichs were right about the boom, but wrong about the bomb. The global population, which stood at 3.5 billion in 1968, has rocketed to 8 billion today. That has undoubtedly caused major environmental stresses, and many of those people are not well nourished. But there has been no global shortage of food. Hunger has certainly persisted, but in localised pockets, and more often as a result of conflict. The worst famine during this period was in Ethiopia from 1983 to 1985, affecting nearly 8 million people and prompting an outpouring of charity, notably the Live Aid concerts in London and Philadelphia. The famine was instigated not by population growth but by drought, and greatly worsened by the country's internal strife. Overall, for the past half-century there's been enough food to go round—as long as it actually *does* go round.

The instructive point isn't that the Ehrlichs got it wrong—any number of people have made doomerist predictions in the past, only to be proven wrong. It's *how* they got it wrong. *The Population Bomb* essentially recapitulated the argument famously made by the English clergyman Thomas Malthus in 1798. In *An Essay on the Principle of Population*, Malthus argued that people can make new people far faster than they can make new arable land. The number of mouths to feed therefore goes up much faster than the amount of food grown for them. Catastrophe follows.

Or doesn't. That catastrophe had *not* followed, on the whole, should have been a clue that Malthus's argument is flawed. In fact, farming was becoming significantly more productive just as he was making his doomy predictions, thanks to a combination of technological and economic innovations: advances in crop rotation, selective breeding and the beginnings of industrial farming and commodity markets. And what refuted the Ehrlichs's neo-Malthusianism 170 years later was the "Green Revolution," a package of agricultural technologies including high-yielding crops, chemical fertilisers, pesticides, irrigation and mechanisation.

The other flaw in the Ehrlichs argument was that while the total population was indeed growing, the *rate* at which it was growing was falling—and continues to fall today. This shouldn't have been unexpected either: it was in line with a pattern noted as early as 1929 by the demographer Warren Thompson, who formulated the demographic transition model we looked at in Chapter 2. It was evident in my own Indian-origin family. My grandparents were small landowners and home-makers, but my parents' generation, including the women, had professional careers. My father was one of ten children, my mother one of five. They had their own children later, and invested more—financially and emotionally—in them. Of those fifteen siblings, two had three kids; all the others had two or fewer. One family's experience can't be truly representative of the whole, of course, but it happened to reflect the trend: population growth, and eventually total population, begins to fall.

Again, the Ehrlichs had explicitly dismissed this possibility. "There are some professional optimists around who like to greet every sign of dropping birth rates with wild pronouncements about the end of the population explosion," they wrote. "They are a little like a person who, after a low temperature of five below zero on December 21, interprets a low of only three below zero on December 22 as a cheery sign of approaching spring." While they acknowledged demographic transition in Taiwan and Korea, they then gave a string of anecdotes illustrating the implausibility of birth control in India, and concluded: "We would be foolish in the extreme to count on similar sequences of events taking place in other parts of Asia, in Africa, or in Latin America." Perhaps they should have had a word with my family.

During one of Paul Ehrlich's *Tonight Show* appearances, he told Johnny Carson that environmental impact was equal to population multiplied by affluence (or consumption) and technology (how many resources are needed). That meant the only way to protect the environment was to reduce population, affluence or technology. But these variables—which are each multi-faceted concepts, not readily expressible as numbers in a single equation—are connected in ways that are much more complicated than straightforward multiplication: affluence turns out to drive population *down*. And technology doesn't have to get more resource-intensive; it can go down, too.

It's impossible to be certain, of course, how things might have worked out if population control measures had not been taken. But that's exactly the point: there are multiple ways things might have played out—multiple scenarios—had different decisions been made about the shape and size of the economy, or about investments in agricultural and other technologies.

While Paul Ehrlich insisted that his book presented potential future scenarios rather than hard predictions, it overwhelmingly focused on a straightforward extrapolation of then-current trends into the future, while dismissing his own equation's implications for the potential effects of technological and social change. Of the multiple stories that could have been told about population in 1968, the Ehrlichs chose to tell only one, and they told it with such vim and vigour as to persuade millions that it was true. They seemed to have produced a robustly "realistic" account. But it was no such thing. It was an exercise in pessimism.

The Poker Game

Physical systems like the weather may be hard to predict, but at least their behaviour doesn't change just because we're *trying* to predict them. Human behaviour, however, does. All it takes to make us act differently is to tell us we're being watched or provide us with new information. Electoral polls can change how people decide to vote; playing the stock market—which in effect means voting with your wallet—is entirely based on the flow of information, including an entire ecosystem of punditry, expectations and forecasts.

My first proper job was at a magazine called *Risk*, where I was charged with editing technical articles about derivatives—financial instruments used for betting on stocks, bonds, currencies and commodities, and more respectably as a kind of insurance against the risks of a company's business activities. Finding the right price for these instruments, and managing their risks, was a formidable task: banks had hired legions of "rocket scientists"—mostly mathematicians and physicists—to build models of how their values would change as time went on and markets moved. It was an expensive business, but not as expensive as getting it wrong. The multi-trillion-dollar derivatives market depended on them.

All of this rested on the work of the French mathematician Louis Bachelier, whose doctoral thesis, published in 1900, first suggested that stock prices moved randomly. That meant they could be modelled using some comparatively simple statistical tools: in particular, price movements should follow a "normal distribution," in which most outcomes are close to the average, a few are fairly distant from it and a tiny number are very distant from it. Plot these out and you get the familiar "bell curve": a low line rising gradually at first, then rapidly to a high hump and then dropping off again symmetrically.

The normal distribution comes up over and over again in nature, and in our everyday lives. Five years after Bachelier's suggestion, for example, Einstein proposed that water molecules move the same way. Human height is normally distributed; so is wind speed, and so are many other scientific variables. That's why it's called the normal distribution. It's business as usual. Nothing to see here.

But Bachelier and his work remained obscure for decades—until they were rediscovered in the 1960s by economists looking for ways to price derivatives. His insight, somewhat modified, became one of the key planks of a model developed by the MIT economists

Fischer Black, Myron Scholes and Robert Merton, and integral to a theory of markets in which rational investors ensure that prices reflect any and all relevant information. Emboldened by this new-found mathematical precision, traders developed a vast array of "exotic" financial instruments.

There was just one problem, as I soon discovered: it was wrong. Prices *don't* move normally, investors are not perfectly rational, and markets are not even remotely efficient. There were far more big swings—when markets soared, and sometimes crashed—than the theory predicted. Of course, I wasn't the first to have noticed this: it had been described years earlier by Benoit Mandelbrot, "the father of fractals," and by the trader and risk theorist Nassim Nicholas Taleb, who later shot to fame with his book *The Black Swan*. Both had published critiques of the industry standard models in *Risk*, but the banks had paid little attention. Crashes happened, everyone acknowledged, but they were just one of those things you had to live with—"Acts of God," like natural disasters. They weren't worth worrying about, particularly for firms that had invested heavily in talented staff and state-of-the-art models.

One such was Long-Term Capital Management (LTCM), a massive hedge fund run by Wall Street legend John Meriwether, whose bold trading style was immortalised in Michael Lewis's book *Liar's Poker*. The fund's models were under the care of Merton and Scholes, who had shared a Nobel prize in 1997. LTCM made money by placing enormous bets, often using complicated derivatives on obscure securities, and borrowing a lot of money to do so. And for a few years it made *a lot* of money. But in the summer of 1998, Russia unexpectedly defaulted on its debt, sending markets into a spin as investors rushed to safety. LTCM stood to lose hundreds of billions of dollars—far more than the value of its assets, and enough to shake the financial system. The Federal Reserve Bank of New York was forced to bang some banks' heads together and organise a rescue package.

What does it mean that a team of such supposedly stellar talents, with all the resources that money can buy and all the incentive in the world to get it right, failed so catastrophically to predict what was coming? In 1994, LTCM had circulated a note to prospective investors which explained that its approach to generating a 25 per cent annual return came with a 1 in 100 chance of *losing* 20 per cent. That was its "worst-case scenario." Now, it claimed, it had experienced a "six-sigma event"—a market movement so extreme as to be expected only once in every 4 *million* years of trading. LTCM had just got stupendously unlucky.

It hadn't. You didn't have to go back 4 million years to find a comparable event, but only a decade: on 19 October 1987, known as "Black Monday," global stock markets had crashed for no readily apparent reason. There were plenty of other examples of massive "unexpected" crashes—including a Russian default just after the Bolsheviks seized power in 1918. LTCM's models, however, were apparently calibrated on just the past few years' worth of data. You might argue that was justifiable, given that the world had been very different in 1987, and still more so in 1918. But it wasn't 4 million years different.

"There was this whole belief that recessions were done, the economic cycle was over, the Cold War was over. It was just going to sort of be happy sailing—I mean, it sounds ridiculous now, but it wasn't ridiculous," said Roger Lowenstein, author of *When Genius Failed*, the definitive account of LTCM's demise. "It was a very seductive point of view after six or seven years of a gradually increasing, accelerating economy, mostly peace on earth, pre-9/11. In their view, which they embodied in their trades, things are going to get better and better and better and better."

If you think this sounds somewhat familiar, you're right. Having had a front-row seat for the LTCM debacle, I hoped the lesson would be learned. It wasn't: policymakers accepted that it had just been "one of those things." A decade later, another one arrived.

The global financial crisis of 2008 wasn't a straightforward repeat, but its roots were similar. This time, instead of a single firm, an entire industry—traders and investors in repackaged mortgage debt—became over-confident about its ability to predict how markets would behave. In this deeply Panglossian collective delusion, financially insecure borrowers could nonetheless get mortgages and live the home-ownership dream; the consequent risks of default could be underwritten, no matter how risky, for a commensurate price; and this financial lead could be transformed into gold through the alchemy of high finance. Everybody wins, in this best of all possible markets.

Or, as it turned out, everybody loses. House prices fell, borrowers couldn't service their mortgages and the whole house of cards came tumbling down. The same excuses were made about the extreme unlikeliness of the events that subsequently tripped the banks up. But just as the unexpected thinning of the Antarctic ice in 2023 reflected how wrong the models were, so did the supposedly unlikely moves in the markets. They weren't a measure of how unlucky the traders had been: they were a measure of how wrong their models were. Nonetheless, the received wisdom in the aftermath of the crash was that nobody could have seen it coming.

That this ultimately didn't stand was thanks to Queen Elizabeth, of all people, who was believed to have lost a considerable chunk of her own fortune in the crash. While opening a new building at the London School of Economics in November 2008, she famously asked *why* no one had seen it coming. That prompted the august British Academy to go in search of a decent answer. Four years later, it eventually reported that many people *had* in fact seen trouble coming, but "most were convinced that banks knew what they were doing. They believed that the financial wizards had found new and clever ways of managing risks . . . It is difficult to recall a greater example of wishful thinking combined with hubris." The debacle, it concluded, "was principally a failure of the collective imagination of many bright people."

Imagination isn't a word generally associated with banking. But after the crisis, it became an integral part of the regulatory framework. With grumpy taxpayers demanding that finance's watchdogs finally step up, banks are now required to take part in regular and stringent "stress tests"—basically, to role-play imaginary scenarios in the form of giant spreadsheets of price and rate movements—and see how their portfolios hold up under the strain. They include various kinds of market upheaval, but also, these days, the effects of climate change and social media–driven financial panics.

It's an odd, numerical kind of imagination, far removed from conjuring up a sun-kissed tropical beach in your mind's eye: but it's envisaging something that's never happened, so it's imagination nonetheless. And that's something we have sometimes forgotten: if you want to predict how the future will turn out, you can't just rely on the models and numbers. Part of the future is visible only in our imaginations.

The Invention of the Future

If you were to pick the one person out of all of humanity's recorded history who most changed how we imagine the future, it would probably have to be Herbert George Wells.

Wells was a socialist and a republican, a eugenicist and, by our standards, a white supremacist.* He wrote copious mainstream fiction exploring the social mores of the day, and non-fiction proposing alternatives; and his science fiction was socially motivated, too. Where his great rival, Jules Verne, wrote "grand adventures"

* Wells expressed sympathy for those ill-treated because of their race, but believed they would eventually vanish or be subsumed into a homogeneous "world state."

detailing how submarines and rockets worked, Wells wrote allegories that explored ethics, morality and progress. His 1895 book, *The Time Machine*, for example, makes no real attempt to explain how the titular machine works; but when the Traveller visits the far future, he discovers it inhabited by the effete Eloi, who enjoy lives of ease and luxury, and brutish Morlocks, who labour underground. These, we are given to understand, are the descendants of the decadent aristocracy and the lumpen proletariat.

That's not to say he wasn't far-sighted on technology, though. Over the following six years, Wells invented almost every major trope of science fiction as we know it today, including genetic engineering (*The Island of Doctor Moreau*, 1896); invisibility (*The Invisible Man*, 1897); alien invasion (*The War of the Worlds*, 1898); suspended animation (*When the Sleeper Wakes*, 1899); and space travel (*The First Men in the Moon*, 1901). In subsequent writings, he also predicted telecommunications, atomic bombs, aircraft, tanks, a universally accessible information resource called a "world brain"—and one of the first novels about a parallel universe, containing a world called Utopia.

Small wonder, perhaps, that he seemed frustrated by the slow pace of change in the here and now. In a 1902 lecture at the Royal Institution, later published as *The Discovery of the Future*, he said: "We travel on roads so narrow that they suffocate our traffic; we live in uncomfortable, inconvenient, life-wasting houses out of a love of familiar shapes and familiar customs and a dread of strangeness; all our public affairs are cramped by local boundaries impossibly restricted and small. Our clothing, our habits of speech, our spelling, our weights and measures, our coinage, our religious and political theories, all witness to the binding power of the past upon our minds."

This, Wells suggested, was because society was dominated by people who clung to the past rather than those who looked forward to the future (or as Philip Zimbardo would put it, "past positives" rather than "future-focused"). That domination, he suggested, was possible because the past seemed more knowable than the future. Perhaps you could say the same about our times: the future seems full of confusion, while the past is soothingly ordered. Or is it? As we saw in Chapter 5, that's something of an illusion: it's just easy to tell a coherent story about the events of the past that masks how much we don't know about them.

But, as Wells went on to say, modern science had made the past more knowable still. Geology, palaeontology and evolutionary biology, albeit still in their infancy, had between them revealed a much deeper and richer past than anyone had expected, and one which had profound consequences for how we humans saw ourselves.

Wells thought we needed to take a similarly rigorous approach to the future. His own work demonstrated that it was possible to make educated guesses about the world-changing technologies ahead, and the opportunities and challenges they might create. The science of prediction was also in its infancy, but its outlines were being drawn up, and while individual humans remained unpredictable, he hoped (like many who followed) that collective behaviour might prove more tractable. Thinking deeply about the ways in which the future might unfold, embracing its possibilities and recognising its perils, might help us to chart a course towards the best outcome. It was a rallying call for optimism: to seek out the best of all possible worlds.

Three decades later he was still banging the drum. "It seems an odd thing to me that though we have thousands and thousands of professors and hundreds of thousands of students of history working upon the records of the past, there's not a single person anywhere who makes a whole-time job of estimating the future consequences of new inventions and new devices," he said during a 1932 radio broadcast on the BBC. "There is not a single Professor of Foresight in the world. But why shouldn't there be? All these new things, these new inventions and new powers, come crowding along; everyone is fraught with consequences, and yet it is only after something has hit us hard that we set about dealing with it."

With hindsight, Wells demonstrated his point pretty well during that same broadcast, whose ostensible subject was the future of communication. In response, he suggested that the impending "abolition of distance" by motorised transport and telecommunications would give rise to two diametrically opposite outcomes. It would allow people around the world to befriend each other as easily as they could the folks next door; but it could just as easily allow death and destruction to be dealt at a remove. As we now know, he was right about both outcomes. But we're only dealing with it now that it's hit us hard.

The Second World War affirmed the need to think ahead in terms of both geopolitics and technology, and birthed outfits like the RAND Corporation, which sought to put futurology on a robust technical—which is to say, military-industrial—base. "Hard" SF writers like Arthur C. Clarke, who prided themselves on their scientifically plausible visions, wrote engineering-heavy futures to match. Buckminster Fuller brought design and environmentalism into the frame in the sixties; Alvin Toffler suggested in his bestselling 1970 book *Future Shock* that the rapid transition to a post-industrial society was too much for mere mortal minds to bear—a sentiment resurrected and reiterated pretty much every time a new breakthrough technology arrives.

Today there are a fair number of Professors of Foresight in the world, and a small army of people who call themselves futurologists, covering everything from trendspotting (will pastels still be in vogue next season?) to technology (will augmented reality be in vogue next year?) to transhumanism (will people still be in vogue next century?).* Some are studious forecasters who make highly specific projections: the demand for semiconductors in ten years' time, for

* Many people who work in this field call themselves "futurists," but that runs the risk of confusion with the Italian art movement of the same name—whose members also had pronounced views on the future. So I prefer futurologist, even if it is an uglier word.

example. Some are creative types who dream up futures as if they were art projects, as they sometimes are. And some are slick think-fluencers who provide corporate entertainment for middle managers.

There is an equally broad range of methodologies on offer, often with esoteric names: visioning, back-casting, scenario planning, speculative design, Kondratieff waves and many, many more. Sometimes these are intended to map out a proposed policy change; sometimes at shoring up a corporate bottom line; sometimes at fending off the collapse of civilisation. This tangle of ideas, disciplines and objectives is confusing, but what much of it shares is that it doesn't aim to forecast the future, but explore its possibilities, using imagery and narrative. You might call it "applied speculation." Or just "telling stories."

It's become a commonplace that stories are powerful, and for good reason: they can convey ideas, arguments and experiences in engaging and memorable ways. Consider the ease with which you can remember, say, a fairy tale or a TED Talk compared with a chapter of a textbook. Stories can endure for thousands of years, because at their best they engage our instincts, or intellects *and* our imaginations. The Pandora myth was first recorded more than 2,500 years ago, but I can still use it to explore the nature of hope today. Now we need to write ourselves new myths, for a brighter future.

The Endless Odyssey

Many of us feel we have little or no control over the world. We might doubt that anyone does. There are many apparently destabilising forces at work, many of them born of the good intentions of generations past. Climate change is a by-product of the Industrial Revolution. Improvements in public health have led to a demographic inversion, or implosion. The socioeconomic regimes that brought prosperity to much of the world in the twentieth century seem to have overshot, leading to unsustainable consumption and extreme inequality. The internet has brought instant access to the world's information and its people, but also social and political instability. New technologies like artificial intelligence or synthetic biology promise great rewards, but might prove to be beyond human control. We hear daily that the future has become wildly unpredictable, from pundits and politicians, business people and clerics, scientists and poets. We see only accidents waiting to happen, and feel no sense of agency. This is a recipe for fatalism, the ingredients of a pessimism trap.

It's important, I think, to recognise that this is not a new state of affairs. The scale and pace of our concerns now may be different to those of previous eras, but imagine the consternation of a Palaeolithic hunter wondering why the game is disappearing, knowing nothing of the impending Ice Age. Or a medieval aristocrat wondering why no amount of money nor medicine could protect them from the plague. Or philosophers trying to explain why the Lisbon earthquake killed so many pious souls. Or those in the 1940s caught up in the war that followed the supposed "war to end all war," as H. G. Wells called it.

In 1937, the French poet Paul Valéry wrote: "The future, like everything else, is no longer quite what it used to be . . . We have lost our traditional methods of thinking about it and forecasting it: that is the pathetic part of the state we are in."* He went on to describe this as discovering, while playing cards, "that his partner's hand contains cards he has never seen before, and that the rules of the game are changed at every throw."

In some respects, we have *less* reason to feel that the future is uncertain than Valéry did in 1937. Aspects of the future that were

* He wasn't the only one. The same year, the American writer Laura Riding also wrote "the future is not what it used to be," in an essay addressed to her lover Robert Graves. People had given up on trying to shape it, she declared, and instead "behave with more and more fatally decisive immediacy."

once entirely unknowable have become mundane. Back then, if you wanted to know if it would rain tomorrow, you would have to consult a barometer, check the prevailing wind and still take an umbrella with you just in case. Today you can be pretty sure that your weather app is going to give you a decent steer. But then again, one of the reasons we believe the future to be so fraught is precisely *because* we've got better at predicting what it holds. Our understanding of the consequences of global warming, patchy though it is, is a miracle of foresight. We can't ignore our predictions and go about our business as if we knew nothing of what was coming. But stressful as that knowledge is to bear, it's better than sleepwalking into disaster.

Forecasting can help us close the gap between our expectations and our realities, but it can only go so far. It works best when we have patterns to go on, but sometimes the pattern becomes broken—or we break it. The key to managing the mix of predictability and change is to tell stories that cover a range of eventualities. Those might be hard-nosed scenarios, driven by data and fitted to mathematical curves, like a weather forecast or a climate model. Or, when we have little hard evidence to go on, they can be imaginative flights of fancy, constructed with an eye to what we want, need and believe. The important thing is that we use them to anticipate and prepare for the many possible challenges of the future, not just the ones that lie most directly ahead of us.

Nature gives us accidents and chaos; we respond with technological innovation and social change. We should take from this that we can make progress; we can harness the processes of the universe, and we can overcome the challenges it presents. This is a philosophical position called "meliorism," described by the psychologist and philosopher William James as a route between Leibniz's optimism and Schopenhauer's pessimism. In his 1907 essay "Pragmatism," he wrote that meliorism "treats salvation as neither inevitable nor impossible. It treats it as a possibility, which becomes more and more of a probability the more numerous the actual conditions of salvation become."

James was talking principally about religious salvation, but later thinkers extended the idea to stand for the belief that it is *possible* to make the world better, through human effort, but progress is by no means guaranteed. Terminology notwithstanding, I think *this* is what it means to practice optimism in the twenty-first century.

In that spirit, I suggest that Oedipus is the wrong myth for our times. For him the future *was* what it used to be—what it always had been, and always would be. We're not omniscient and certainly not omnipotent, but our capabilities are such that the likes of Sophocles or Aristotle might have *considered* them to be godlike. And we are inventing new ones all the time. We can find a better model in a different myth—one older than Oedipus, and in fact one of the oldest we have.*

Odysseus was the shrewd military advisor who devised the Trojan Horse, and thereby won the Trojan War for the Greeks. But, as related in the *Odyssey*, on his way home to Ithaca he is blown off-course and captured by the one-eyed giant Polyphemus, son of the ocean god, Poseidon. Having got Polyphemus drunk, Odysseus blinds him; he and his men escape by hanging on to the bellies of the giant's sheep as they go out to graze.

His victory is short-lived, because Polyphemus calls on his divine father to blight Odysseus's journey home; from then on, the hero's return is impeded by one thing after another. The witch Circe attempts to ensnare him, but he resists her magic by taking a sacred drug; he lashes himself to his ship's mast to avoid being lured to his death by the song of the Sirens; navigates the perilous race between

* The anthropologist Julien d'Huy has suggested that the story of Odysseus and the cyclops Polyphemus is a version of an ancestral story that goes back 18,000 years.

the six-armed monster Scylla and the whirlpool Charybdis; and is once again trapped, this time by the nymph Calypso, who offers him immortality, which he refuses. He just wants to get home.

When he finally escapes Calypso and returns home, twenty years after leaving it, he finds his house full of suitors theoretically vying for his wife's hand in marriage, but in fact living it up off his estate. Disguised as a beggar, he infiltrates a contest intended to disqualify all the suitors—the task being to string his own bow. Having done so, slaughtered the suitors and proved his identity to Penelope, he finally settles down to a well-earned rest. What he finds upon his return home is that some things have changed, and some have stayed the same. Like all nostalgics, both ancient and modern, he learns that you can never truly go back.

Our journey into the future is likewise punctuated by incidents and accidents; sometimes we have warning that they are coming, and sometimes we don't. To overcome them, we have to predict and prepare for them as best we can, using all the wit and imagination we can extend: the challenges continue until we finally reach our destination; and when we reach it, we will find new challenges to overcome.

In the opening lines of the *Odyssey*, the hero is introduced to us as *polytropos*, a man with many turns. He rises to his succession of trials using a combination of reason and creativity, intellect and intuition. That inspired Murray Gell-Mann—best known as a physicist but himself a man of many turns, from archaeology to arms control—to declare Odysseus as the model to whom we should aspire in the modern age: neither purely analytical, nor wholly intuitive, but combining both to tackle the world's wicked problems.

Not many of us possess such a combination of attributes individually: we can't all be Gell-Mann. But collectively, we do. The future has many authors, and so we have to predict, imagine and act, together. We can make Leibnizian grand designs—attempting to understand and reshape the world in our own image—or we can garden, tending only to that part of the world we can ourselves reach. If we want to make a better future—the best of all possible futures—we will need to use our superpower of mental time travel to the full, to imagine what that future might be like. And then we can set about changing it. Or rather, creating it.

Nor is there just *one* future. There are innumerable worlds we can visualise, and many we can realise through our actions. Our collective task is to find the best of those possible worlds, using our skill, knowledge and judgment. And since what happens in the future, or futures, is linked to what we do in the present, we have to act now to bring it about. That means we need to stop trying to understand optimism, and start getting on with it. It means taking optimism out into the world.

PART III

Optimism in the World: Potential Futures

How do optimists change the world—for them and for us?

Are things getting better, worse—or neither?

What do we do about challenges we've never encountered before?

7

Taming Panglossians

How do optimists change the world—for them and for us?

As the twentieth century drew to its close, jokes about the impending end of the world became common. Sometimes they were loaded with a touch of superstition: obviously the turn of the millennium was just a date like any other . . . but what if, just maybe, it was something more? Like most people, I didn't really expect the world to end at the stroke of midnight on 31 December 1999. But I did wonder if a few planes might fall from the sky.

The last day of the twentieth century and of the second millennium* was crunch time for the so-called "Y2K bug": the inability of many older computers to process dates beginning with 20 instead of 19. The fear was that this might cause critical computer systems to fail, bringing down everything from banking to air traffic control. The hope was that catastrophe could be prevented by a sufficiently coordinated and determined effort.

"If we act properly, we won't look back on this as a headache, sort of the last failed challenge of the twentieth century," US president Bill Clinton had said in 1998. "It will be the first challenge of the twenty-first century successfully met." We did act properly: a global

* Yes, that *actually* came a year later. But you tell that to the people dancing in the streets.

effort to patch and upgrade vulnerable systems followed. Nothing major collapsed; certainly no planes fell from the sky. The first challenge of the twenty-first century had indeed been successfully met. After the trials of the twentieth century, the new millennium seemed to be off to a bright start.

Twenty months and eleven days later, Islamist terrorists flew passenger jets into each of the twin towers of the World Trade Center in New York, and a third into the Pentagon in Washington DC. A fourth flew into the ground in Pennsylvania, after passengers fought back. Almost 3,000 people were killed. The millennial euphoria evaporated. The future suddenly looked dark.

I'd spent most of 1997 in New York, setting up a local office for *Risk*, the magazine where I'd learned about high finance; and had spent a week or two in Manhattan every month in the years preceding 9/11, as required by my next gig as the editor of a start-up online news service. I had friends there, regular spots, favourite views, routine walks. I'd taken many visitors to the viewing deck on top of the WTC; it was one of Kathryn's favourite viewpoints in the city. I considered it a bit brash; I preferred the dingy but atmospheric Empire State Building, which loomed over my Murray Hill apartment.

On the morning of the attacks, however, I was back home in London, where I watched the live coverage on my tiny kitchen TV set, overcome by a numbing sense of unreality. It barely occurred to me, until the moment the South Tower collapsed, that the remaining occupants were going to die; I assumed they would somehow be rescued. Unbeknownst to me, those occupants included the staff and delegates at an event being held on the 110th floor of the North Tower by Risk Publications. I found out afterwards that many of its staff, including a number of people I'd worked with, been to the pub with, grown up with, had been killed. A couple planning to get married; my former office manager, a moonlighting actor and part-time dog-walker; a former colleague I'd failed to recruit to my new business.

I didn't go back to New York for months. After that, not for years.

After the first few tense months it became evident that plane travel wasn't much riskier than it had been before, although inchoate anxieties remained: bombs in the hold, mechanical sabotage, surface-to-air missiles. It was clear that the chances of disaster remained tiny on any given flight . . . but what if, just maybe? I didn't have a lot of faith in the "security theatre" being performed at airports. Air travel had become a tense and unpleasant experience; and at the end of the flight would be a city I thought of as my second home, plunged into mourning. Once I'd visited to pay my respects at the WTC site, I didn't want to go back again—particularly when, a few years later, I was plunged into mourning of my own.

I was not alone. Many people stopped travelling, particularly by air, after the attacks. Harvard psychologist Steven Pinker thought this was overdoing it. "The greatest damage that resulted from the attacks was self-inflicted, in our individual and national overreactions to them," he told Julia Belluz of *Vox* in 2016. By way of evidence, Pinker cited analysis conducted by German psychologist Gerd Gigerenzer which shows that "after 9/11, 1,500 Americans died in car accidents because they chose to drive rather than fly, unaware that a car trip of twelve miles has the same risk of death as a plane trip of three thousand miles."*

The 9/11 attacks, spectacular though they were, didn't make any significant difference to the risks of flying, working in a tall building or being killed by terrorists. More Americans drowned that year, or were killed in motorcycle accidents, than were killed by terrorists on 9/11—deaths considered tragic but unremarkable, and which certainly didn't prompt a massive rethink of attitudes to safety, much less the expenditure of billions of dollars in precautionary measures. In 2024, the arch-rationalist blogger Scott Alexander wrote that, had we investigated the statistics of terror

* This has been disputed—a subsequent and seemingly more careful analysis found that deaths in car crashes did not rise, although "non-severe injuries" did.

attacks, "we should have shrugged, said 'Yup, there's the once-per-fifty-years enormous attack, right on schedule,' and continued the same amount of counter-terrorism funding as always."

That, of course, was *not* how the world reacted. "If to this day we have a clear memory of where we were when we heard of the attack, or how we felt and other circumstances, it's because, deep down, we realised then that the world had changed," wrote the Portuguese writer and politician Rui Tavares in *A Short Book on the Great Earthquake*. Just as the destruction of Lisbon in 1755 forced people to reappraise their expectations of the world, so did 9/11. The West realised that all its wealth and might could not shield it from unconventional assaults by determined enemies. So did those enemies.

Perhaps the relative stability of the eighties and nineties had blinded us to the potential for world-changing events. If you'd lived through the first half of the twentieth century, two world wars and a Great Depression would have left you in little doubt that sharp changes of fortune are a fact of life. But in the second half of the century, it became easier to ignore such unpleasantness. The Cold War stand-off between the superpowers might have meant a certain level of existential dread, but it meant that fewer wars got hot. Optimism was comprised, perhaps, of the expectation that this state of affairs would continue indefinitely.

And starting in the fifties, the combination of technological innovation, liberal democracy and laissez-faire economics consistently delivered improved standards of living. By the nineties, this process seemed to have reached the level of destiny, as heralded in political scientist Francis Fukuyama's 1992 book, *The End of History and the Last Man*. Fukuyama had been inspired by the fall of the Berlin Wall and the collapse of the Soviet Union to take the Hegelian position that civilisation had identified its ideal form—and contra Marx, it was not communism. With China and India opening up, too, a clear majority of the world's population seemed to be heading towards freedom of belief, trade and movement. The 9/11 attacks threw this rationale for geopolitical optimism into profound doubt.

Three years later, a huge tsunami swept across the Indian Ocean, killing nearly a quarter of a million people and causing vast amounts of damage. On the face of it, it had nothing in common with 9/11 except that it was an unexpected tragedy. But Tavares argued that a thread connected the two events. "The world we see is the world we have," he wrote. "They were experienced by the same generations, on a worldwide scale, in a very similar way across the world." He cited the emphasis placed on the religious, ethnic and social diversity of those killed on Boxing Day 2004; that was now considered particularly relevant, given the divisions opened up by the events of September 2001.

Tavares's monograph was published in 2005, so makes no reference to the further shocks that were to come: the financial crisis, Covid, the Russia–Ukraine War. But you could extend his argument to say that our perceptions of those events, too, have been coloured by what has gone before; all the more so since we live in a time when you can revisit them with a few taps of a finger—or participate in them vicariously as they arise. Perhaps the continual, immersive nature of our engagement with these disasters distinguishes our times, too. In 1755, 1816 or even 1935, we might have had months or years to assimilate such events; we might well have been oblivious to them, just as Shelley and Byron were to the Tamboran eruption. And while some risks have been tamed, others have been newly recognised or invented, from asteroid strikes to runaway artificial intelligence.

Add all this together and the future seems a mess of uncertainty. Everywhere we turn—and so often, it's to a screen in front of our face—we're confronted with crises and catastrophes. Every strand seems intertwined with every other: a war in Ukraine speaks to the fragility of everyday life, but also to supply chains, energy economics, political despotism, climate change and food security. It's not just a crisis: it's a polycrisis. Or a permacrisis. Perhaps polycrises. Singular or plural? We can't even agree on the terminology.

Under these circumstances, it's easy to see why we might remain upbeat about our own lives—where we can fall back on our innate optimism and sense of control—but struggle to take a similarly positive view of the world in general. We might look to our leaders to show us the way, to find the optimism we need. But all too often, we find it serves them better than it serves us.

Propagandists

It was a presidential entrance like no other. On 1 May 2003, President George W. Bush arrived on the deck of the USS *Abraham Lincoln* in the passenger seat of a Lockheed S-3 Viking jet. Having greeted the top brass and paused for photographs with excited crew members, he changed into a dark business suit and took to a podium in front of a large Stars and Stripes banner.

MISSION ACCOMPLISHED, it read.

The *Abraham Lincoln* had just returned home from the Persian Gulf, where it had played a key role in the US-led invasion of Iraq, presented as the ultimate response to the 9/11 attacks. Iraq had not been involved in those attacks, but Bush had successfully persuaded the American public that the country's dictatorial president, Saddam Hussein, was amassing "weapons of mass destruction" that could lead to even greater devastation. His removal from power was a matter of urgency. Now that job was done, Bush told his audience—both onboard and at home. "Major combat operations in Iraq have ended," he said. "In the battle of Iraq, the United States and our allies have prevailed."

As it turned out, they had not.

"I've learned that people want to follow an optimist. They don't respond to the message: 'Follow me, things are going to get worse,'" Bush had said when announcing his candidacy. "They respond to someone who appeals to our better angels, not our darker impulses. They respond to someone who sees better times—and I see better times."

However sincere Bush was in his optimism, his promised better times did not manifest. Perhaps when he stood on the deck of the *Abraham Lincoln* he was hoping to reset the American people's expectations, taking them back to the mood before 9/11. But that was not to be. While "major combat operations" might have come to an end, the Iraq War was only beginning. When he gave his speech, the US had lost just over a hundred soldiers in Iraq. By the time military activity *actually* ceased, in 2011, around 4,500 American soldiers had been killed, and hundreds of thousands of Iraqi civilians. Countless more had been wounded or made destitute. The country was in disarray. And three years later, the US was back, helping the reconstituted Iraqi Army put down Islamic State's insurrection. This time round, victory really was achieved fairly efficiently; but there were no showy declarations of victory.

Bush's adventure in Iraq stands in stark contrast to the careful planning that underpins most successful military strategy. Looking back, it's astonishing how little thought the architects of "Operation Iraqi Freedom" seemed to have given to any outcome but their preferred one. Strategists drew up colour-coded maps of neatly redrawn boundaries across the Middle East, just as counterfactual hobbyists might. Pundits described how self-determination in Iraq would sow the seeds of freedom across the Fertile Crescent. They simply didn't give any serious thought to the possibility that things might turn out any other way.

In this, they may have been influenced by Fukuyama's book, much read by the neoconservatives who supported and staffed the Bush administration. (Fukuyama himself was not in favour of the war and eventually parted ways with the neocons over it.) In this scheme of things, overthrowing Middle Eastern autocrats would inevitably lead to the rise of freer societies. Coupled with Bush's belief that he was doing God's work, and overweening confidence in America's status as the world's solitary "hyperpower," there seemed to be no way things could go wrong.

As an anonymous presidential aide (widely assumed to be Karl Rove) told the *New York Times*: "We're an empire now, and when we act, we create our own reality. And while you're studying that reality . . . we'll act again, creating other new realities, which you can study too, and that's how things will sort out." Or, to put it another way: with each sure-footed move, the US would create a new world—not just a new world, but, by its estimation, the *best* of all possible worlds. The power that Leibniz had reserved for God when creating the world had been appropriated by US neoconservatives intent on remaking it. But perhaps they should have been reading Voltaire, not Fukuyama.

Panglossians

Doctor Pangloss was an unfair caricature of Gottfried Leibniz, whose formulation of optimism was more sophisticated than Voltaire acknowledged in *Candide*. But the world is nonetheless full of Panglossians, who behave as though "all is for the best, in this best of all possible worlds." No matter what happens, it'll work out fine.

Once you start looking, Panglossianism is more widespread than you might at first imagine. Most of us have practiced a mild version of it at one time or another. Maybe you've told a friend it was actually a blessing that they didn't get that job, or that a relationship didn't work out. This is rationalisation by counterfactual: if you *had* got that job, or hadn't broken up, you'd definitely be miserable: no other possibilities exist. Even if no one really believes that to be true, it can be more comforting than confronting the bald truth of failure. (How many times have you said, "It's probably for the best"?)

We've seen how optimism can be a valuable quality for a leader: people gravitate to optimists, and public shows of optimism can become self-fulfilling prophecies. Those arguments seem to be borne out by experience—to a point. George W. Bush was far from the first American president to present himself as an out-and-proud optimist. Martin Seligman, the originator of learned optimism, and his protégé Harold Zullow analysed all the speeches given by US presidential candidates between 1900 and 1984, looking for examples of pessimistic explanatory style; their conclusion was that the more optimistic orator almost always won.

The contrast was particularly distinct in the 1980 campaign between the incumbent Democrat, Jimmy Carter, and his Republican rival Ronald Reagan. In July 1979, Carter—struggling with a "stagflating" economy and an energy crisis—had taken to the airwaves to broadcast what became known as his "malaise speech." In it, he suggested that the real problem was rooted not only in policy and governance failures, but in values. "The erosion of our confidence in the future is threatening to destroy the social and the political fabric of America," he warned. Confidence in the future. Optimism.

The immediate reaction was positive: Carter's approval rating leapt. But over time, the conclusion seemed to be that he was talking the talk rather than walking the walk, particularly as the country's troubles mounted further. When Ronald Reagan emerged as his challenger—campaigning under the sunny slogan "Let's Make America Great Again"—voters saw someone whose optimism they believed in. Reagan subsequently won re-election with a similarly optimistic message: "It's Morning in America."

Positivity became part of the political playbook. "Every presidential candidate for thirty years has sought the optimist label,"

Seligman wrote in 2016. "Bob Dole proclaimed himself to be 'the most optimistic man in America,' while Bill Clinton promoted himself as the man from Hope. Mitt Romney went with the slogan 'a better future,' while Obama was the candidate of hope, change and 'yes, we can.' John Kerry told us that 'there is nothing more pessimistic than saying America can't do better. We can do better and we will. We're the optimists.'"

The problem was that these later declarations of optimism said little that was reliable about who went on to win, as the list above reflects. Seligman's methodology, too, produced mixed results; his explanation was that campaign speeches had become so carefully scripted that authentic optimism was becoming harder to detect—for voters as well as psychologists. Perhaps leaders do sometimes profess optimism that they don't really feel, but this might also be a case of faking it until you make it. As with social media influencers, looking and sounding successful can be key to winning support. For politicians, things often *do* simply work out for the best, as far as they're concerned: kindly benefactors wine and dine you, treat you to all-expenses-paid vacations and even take care of your money problems.

Ersatz optimism has its advantages, politically speaking. In an era of soundbite and social media politics, a more nuanced approach is unlikely to "cut through." Suggesting alternatives to the current political trajectory might sometimes be necessary, but is politically risky. And above all, Panglossianism doesn't require us to engage with the complexities of reality. After all, who *doesn't* want to believe that a war has been won, or that there's nothing really to be done about global poverty, or that climate change isn't really a problem, or that we don't *really* need to take precautions against Covid-19?

This tendency reached its zenith, of course, with the triumph of the populists in 2016. Donald Trump's speeches might have been full of American carnage, but he also promised to Make America

Great Again, to an audience convinced that he was himself a Great American success story. "I could stand in the middle of 5th Avenue and shoot somebody and I wouldn't lose voters," Trump notoriously told supporters at a January 2016 rally in Iowa—an unrealistic expectation if ever there was one, but one that seemed ever closer to the truth as his campaign went on. Trump's own explanatory style ticks Seligman's boxes: everything that afflicts him, or ails America, is caused by others, has simple causes with simple solutions, and could be fixed in a jiffy. But only if you vote for him. Or, even better, donate.

It's not surprising that leaders like Trump might tend to the Panglossian, particularly during times of turmoil. For those who have inherited fortunes, Panglossianism comes naturally: with few worldly worries, careers that consist of failing upwards and endlessly generous benefactors, things often *do* simply work out for the best, as far as they're concerned. In the run of things, such blind optimism might be fine. But personal peccadilloes can balloon into real problems when the person deploying them is in a position of power, because the best of all possible worlds for the rich and powerful is not necessarily the best for the rest of us. And the rich and powerful can mobilise their wealth and influence to ignore or stymie remedies for injustice or misfortune, even as they rationalise those injustices and misfortunes as simply the way of the world.

The power of optimism cuts both ways: it can be used to reassure and inspire—yes, we can!—but also to instil complacency in the face of real and pressing challenges. When misfortune eventually arrives at your doorstep, your inaction may be seen not as confidence but as folly. Repeating that "all is well," even as the evidence stacks up that it isn't, weakens your hold. But reality can only be delayed for so long. The future, with its payload of consequences, will insist on arriving—no matter how rich or powerful you are.

Pandemics

St Thomas' Hospital is just a short walk from 10 Downing Street, across the River Thames. But on Sunday, 5 April 2020, Boris Johnson made the trip by car, because—despite official denials—he was struggling to breathe. The Covid-stricken UK watched incredulously as its prime minister was first admitted, then moved to intensive care. When he emerged after a few tense days, he seemed, for once, subdued.

Johnson was admitted a little more than a fortnight after he'd told the nation that normality would be restored in twelve weeks. "I'm absolutely confident that we can send coronavirus packing in this country," he said. At that time, he was resisting advice to put the country into lockdown: according to later testimony, he didn't seem to believe it could possibly be necessary. Other, less dramatic measures like mass testing and social distancing could do the job, he insisted. While he eventually did order the country to be locked down, the pattern recurred with successive waves of the virus. According to later testimony, and armchair psychiatry, he seemed to struggle to reconcile what needed to be done with his sunny persona. His optimism may have cost thousands, or tens of thousands of lives.

Johnson wasn't the only Panglossian populist who was slow to act on the coronavirus. "One day it's like a miracle, it will disappear," Donald Trump told the press in February 2020. In March, Brazil's Jair Bolsonaro opined: "In my understanding, the destructive power of this virus is overestimated." All three politicians made a show of continuing their meeting and greeting as the pandemic took hold, and all three flouted infection control measures. And all three contracted serious cases of the virus.

This kind of behaviour seems to have been motivated, in part, by the belief that the public needed shows of confidence—the kind of self-fulfilling prophecy that serves well under other circumstances. "We just need more optimism," concluded arch-sceptic Senator

Rand Paul after a June 2020 showdown with Anthony Fauci, in which Paul suggested that there was room to be hopeful that Covid would have a mild winter season.

But the public did not, in fact, want that. Near the beginning of the pandemic, Vanderbilt University psychologist Jane Miller and colleagues asked several hundred participants whether they thought other people should be optimistic, realistic or pessimistic when estimating the risk of various Covid-related problems. What they found was that "people did not tend to prescribe optimism for the dangers. They did not prescribe it for anyone—not for themselves, friends, family, the average person, policy makers, or politicians. In fact, for estimating the likelihoods of three of the most critical and immediate outcomes discussed at the time—getting COVID, being hospitalized, and experiencing a respirator shortage—most people prescribed pessimism." The leaders' professed optimism—which might well have been genuine, given their cavalier behaviour—was running way ahead of the citizens they represented.

Error management seemed to be key. The participants in Miller's survey were aware that they were likely to wrongly estimate the risks associated with Covid: being too optimistic and catching an untreatable virus posed a much greater danger than being too pessimistic and not catching it. It was the opposite of the situation faced by the mouse deciding whether to go out foraging. Perhaps that also helps explain why compliance with lockdown was so high in the early days of Covid—much higher than expected by some in government—and why it tapered off later on when the risks appeared to be better understood.

Miller's interest, incidentally, stemmed from Covid's status as a novel crisis—of the sort that might increasingly stem from climate change, her principal object of study. A Panglossian approach from leaders might again be out of kilter with the citizenry; that's supported by opinion polls which show considerably more public support for significant action on climate change than is reflected by political actions.

But while Panglossian optimism might have cost lives during the crisis, a more Leibnizian form saved them. Although the onset of Covid was dizzyingly rapid, a novel pandemic had been at the top of the UK government's Risk Register for a *decade*. There was ample historical evidence, from the medieval Black Death, the 1918 flu pandemic following World War I and the lived experience of the 2003 SARS and 2012 MERS coronavirus outbreaks, as well as a (thankfully mild) swine flu pandemic in 2009 which demonstrated just how quickly disease could now travel, undetected, around the world. Epidemiologists had been warning for years that a new pandemic was just a matter of time. And they had worked through the possible worlds in which it had arrived. Many of those worlds existed only inside a computer: as models of the nature and behaviour of a novel virus—its physical distribution, its patterns of mutation and transmission, its effect on various more or less vulnerable sectors of the population. But there were more tangible simulations, too.

In 2016, the UK government conducted Exercise Cygnus, a three-day simulation in which nearly a thousand people from central and local government, health services and emergency services worked through their response to a fictional "swan flu" virus. The result anticipated many problems that later arose, such as the difficulty of managing the disease in care homes for the elderly, and flagged that the health service was likely to collapse if the disease was allowed to spread unchecked, suggesting lockdowns would be needed. That same year, the UK also conducted the smaller Exercise Alice to figure out how the UK could respond to an outbreak of MERS, the coronavirus which had already caused lethal, but contained, outbreaks in South Korea and Saudi Arabia. This flagged the need for protective equipment for healthcare workers, effective contact tracing and a rapid ban on travel from overseas.

Covid, then, was neither truly unexpected, and nor were countries totally unprepared. Despite the claim made in thousands of blathering LinkedIn posts and Powerpoint presentations, Covid wasn't a "black swan" event. Nicholas Taleb, the risk theorist who coined that term, has expressed irritation at its use to describe the pandemic, since it was in his view (and mine) wholly foreseeable. "If [the US federal government] can spend trillions stockpiling nuclear weapons, it ought to spend tens of billions stockpiling ventilators and testing kits," Taleb said just after Covid broke out.

As it was, such stockpiles were not amassed; in the UK, stocks of protective equipment were wound *down* for no particular reason except thrift, leaving medical staff exposed and requiring vast expense to hastily acquire when Covid broke out. Nor was there any investment in capabilities like contact tracing or rapid vaccine production. It's no coincidence that some of the countries with the most effective initial reactions to Covid were in Southeast Asia, which had lived through the starkly terrifying experience of SARS and MERS, coronaviruses far more lethal than Covid: they were equipped and ready to adopt masks, measure temperatures and institute lockdowns.

Johnson and his government got a lot wrong—a *lot*—but one instance in which their optimism was justified was their belief that a safe and effective vaccination would be developed, manufactured and deployed at a pace that would have been entirely unimaginable just a few years before. Initial estimates, based on previous experience, suggested it might be up to a decade before a vaccine was available. But in fact, it took less than a year to get the first arm jabbed with the vaccine, thanks to the novel drug development technology of messenger RNA (mRNA), copious funding and an accelerated trials regimen.

That, too, was an example of optimism. Katalin Karikó won the Nobel Prize for Medicine, along with her collaborator Drew Weissman, for her work in developing the vaccine—which, to many people, must've seemed to come out of nowhere. In fact, it was the result of decades of work, most of it in the face of utter disinterest.

Karikó, whose family lived in a one-room apartment in central Hungary, climbed up the educational system, eventually gaining a doctorate in biology, with a focus on mRNA, before moving to the US. Sadly, no one there—except fellow researcher Weissman—was interested in mRNA, a delicate molecule that didn't last long in the body and wasn't viewed as useful to medicine. Karikó was known as "that crazy mRNA lady," labouring away at a project no one expected to work out and unable to secure funding. Eventually, her university threw her out on the street—literally, dumping her equipment outside the building. But fortunately, she was hired by BioNTech, the forward-looking immunotherapy company that eventually collaborated with Pfizer on its Covid vaccine. That was just a start: mRNA technology now looks set to treat many other conditions, including dengue fever, malaria, tuberculosis, Ebola—and cancer.

Was it luck that such a revolutionary technology became available just when the world needed it most? Yes. But on the other hand, it was also testament to the unflagging personal optimism of Katalin Karikó, and to the institutional optimism of the organisations who backed her research in the "unrealistic" expectation that it might just pay off one day.

"I have always believed that scientific research is another domain where a form of optimism is essential to success," wrote Daniel Kahneman in *Thinking Fast and Slow*. "I have yet to meet a successful scientist who lacks the ability to exaggerate the importance of what he or she is doing, and I believe that someone who lacks a delusional sense of significance will wilt in the face of repeated experiences of multiple small failures and rare successes, the fate of most researchers."

In a way, this is also a story of possible worlds: basic research, often into unprepossessing areas, can proliferate into a multitude of applications—just as research into subatomic physics in the early twentieth century gave us the transistor, and from that, the many,

inherited a stately home in Northumberland and a seat in the House of Lords, he also followed his father into the chairmanship of the UK's Northern Rock bank. An Old Etonian turned journalist (like Boris Johnson), Ridley produced several accomplished books on biology and genetics before turning to screeds about how the world is getting steadily better and will continue to do so if we just let laissez-faire economic policy do its thing, most notably in his 2010 book, *The Rational Optimist*. There wasn't much laissez faire in evidence, however, when in the autumn of 2007, Northern Rock became the subject of the first bank run in the UK since 1878, leading to its emergency nationalisation a few months later.

Ridley is at odds with the scientific mainstream in several areas, notably on climate change. A self-declared "lukewarmer," he nods to the reality of climate change but has argued that its effects will be modest or even beneficial. For example, he has claimed that "global greening"—the increase in vegetation due to increased carbon dioxide levels in the atmosphere—will be beneficial for agriculture. Global greening is real, but few scientists—including the authors of one of the papers on which his claim rests—believe there are grounds for calling it a net positive for the planet or its inhabitants, since it represents ecosystems being thrown out of balance, not just more lovely leafy plants. More recently, he has become influential in promoting the idea that Covid originated in a Chinese lab, not in wild animals.

My colleague Liz Else wrote about *The Rational Optimist* for *New Scientist*, sending a randomly selected extract to experts in the field for informal review. The field in question was ocean acidification, a key consequence of global warming. Their judgment of Ridley's assessment was scathing. "The man does not understand the differences in ocean carbonate chemistry controls on short and long timescales, and he compares apples and eggs" was one of the politer responses. "I think it is extremely unfortunate that Matt Ridley has missed many of the important points and concepts. In my view, he has also cherry-picked evidence to form opinions which are unsupported by the bulk of scientific evidence and understanding," read one of the blunter ones. "This is demonstrated by the fact that he completely ignores the mainstream scientific literature. In my view, it is also clear that he has a very poor understanding of the core issues." None of the reviews were rude enough, however, to adduce that Ridley's views might relate to his ownership of working coal mines on his Northumberland estate. Ridley, for his part, promptly wrote a rebuttal, citing more studies to back up his point. After all, he was just playing the Glad Game. And people like him never lose at it.

Ridley is upfront about his stake in fossil fuels, writing on his website that he is "proud that the coal mining on my land contributes to the local and national economy." By arguing for the development of gas reserves, he continues, "I consistently argue against my own financial interest." But he is also an advisor to the Global Warming Policy Foundation, which once propagated old-school climate denialism of the variety that insists warming isn't happening or, if it is, people have nothing to do with it. Now the effects of climate change, from melting ice to rampant wildfires, have become all but undeniable, except by an ever-dwindling handful of lunatics. So the GWPF has moved on to the seemingly softer position that, actually, we're better off adapting to global warming than trying to prevent it—and that it has its upsides.

GWPF trustee David Frost took it upon himself to announce, during the middle of the world's hottest-ever summer, that rising temperatures were "likely to be beneficial" in the UK, because fewer people would die of cold during the winter even if a few more died during the summer. That large areas of the world will likely become uninhabitable if climate change runs unchecked did not figure in his argument that we should move away from renewables or lifestyle changes and "pursue mitigation in a different way." What exactly that "different way" might be was unclear, but one might guess that it will be remarkably similar to "business as usual." Climate change *does* have its upsides, for some people, in some places. I enjoy the dramatic increase in English wine's palatability, for one. But setting these rewards against the massive risks to the Earth's environmental integrity is playing the Glad Game *in excelsis*.

Frost, incidentally, was the unelected advisor whose masterful negotiations finally secured Brexit for Boris Johnson; he played the Glad Game there, too, with the insistence that the minute gains made since Britain's departure from the European Union represented a triumph of national destiny. And the GWPF is one of a cluster of conservative think-tanks located at 55 Tufton Street in London, a locus of dark-money influence in British politics; its housemates were among the architects of the Eurosceptics' triumph.

Dark money also works to sow doubt in climate mitigation at scale—for example, by funding slickly produced YouTube videos that minimise the downsides of climate change. These videos find reasons to be cheerful about global warming—in the same way that Ridley was cheerful about global greening, or Frost about warmer winters—while emphasising the upfront economic and social costs of climate solutions. The long-term benefits of such solutions—healthier environments, new industries and jobs—go unmentioned, while the enduring damage caused by global warming is shrugged off.

In this case, Panglossian optimism is being used against us. We'd *like* to believe climate change is no big deal, and therefore doesn't require us to make any lifestyle changes. It's easier to imagine the status quo than it is the dramatically different world that would exist if we aimed for zero emissions, sustainable consumption and, incidentally, a massive transfer of cash away from fossil fuel companies and their lobbyists. The invitation to play the Glad Game is tempting. But we should resist.

Plutocrats

"One of my favourite sayings is that there are two kinds of people in the world: optimists and pessimists. And the saying goes that optimists tend to be successful and pessimists tend to be right," Mark Zuckerberg told a roomful of do-gooders at Facebook's Social Good Forum in November 2017. "The idea is that if you think something is going to be terrible and it is going to fail, then you are going to look for the data points that prove you right and you will find them. That is what pessimists do."

Zuckerberg had a point. Optimism has certainly helped many people become amazingly successful. In business, as in other walks of life, it encourages risk-taking in the hope of finding previously unrecognised opportunities: which is to say, entrepreneurship. "Optimism is an essential ingredient of innovation. How else can the individual welcome change over security?" asked Robert Noyce, co-founder of Intel and co-inventor of the integrated circuit, the hardware component critical to modern computing.

While entrepreneurs are reasonably good at estimating the chances of a business succeeding on average, they tend to dramatically overstate their *own* chances of success. In a 1988 survey, four-fifths of entrepreneurs claimed their business had more than a 70 per cent chance of surviving; a third said there was absolutely *no* chance they would fail; they were equally bullish on key performance indicators like sales and staff numbers.

You might think they would be kept in line by investors, who put prospective business plans under the microscope. But if the offering is entirely new, all *anyone* has to go on is promises, guesses and hope. Investors in start-ups are well aware of this: they are fond of saying they back founders, not businesses. Optimists, as we've seen, are good at coping with setbacks and bouncing back: a suitably tenacious, capable and motivated founder will *find* a way to make a business work—possibly in ways that weren't envisaged at the outset. Some of Silicon Valley's biggest and most influential companies started out as something quite different. Stewart Butterfield co-founded both Flickr, which pioneered online photo-sharing, and the team messaging platform Slack. In both cases, he had set out to make an online game, but the internal communication tools that his teams developed proved to be the real hits.

Because optimists are better at coping with adversity, believe they possess superior abilities, deal better with stress and pay more attention to good news than bad, they are more likely to press on with their ideas. Even complete failure doesn't put truly optimistic entrepreneurs off: they remain positive as they step away from the smouldering embers of a failed venture and embark on the next one. Optimists find new opportunities where pessimists see only failure; their boosterish initial claims can turn out to be self-fulfilling prophecies. Optimism is an investable proposition.

Nowhere has this principle been embraced more eagerly than in the start-ups beloved of Silicon Valley and its imitators. Online businesses can reach millions or even billions of people without investing very much at all in staff or assets. Downloads on Napster, the early 2000s file-sharing service started by Shawn Fanning while still a student at Northeastern University, outstripped the entire music industry within a few months of its launch; its peer-to-peer system meant it didn't have to pay for servers or bandwidth. The record companies eventually saw Napster off, but the template had been set: it became completely acceptable to make extraordinary claims about disrupting entire industries with next to no hard assets.

Such claims often proved to be wildly unrealistic—except for the few founders that they made wildly wealthy. They can help the rest of us, too, because they propel those founders, and their organisations, to develop products and services that might take much longer to develop under more level-headed regimes. Who would have expected that an encyclopaedia written and maintained entirely by volunteers would become one of the world's most compendious and visited websites? Wikipedia was built on pure, uncut optimism.

But there's a limit. Business is, by nature, brutally indifferent to wishful thinking, and entrepreneurial failure is extremely common. At some point, for the majority of start-ups, self-belief proves inadequate to bridge the yawning gulf between real challenges and actual capabilities, and extreme optimism is disproportionately likely to end in failure: whether by flaming out before achieving take-off, or by gradually dropping back to Earth. The ultra-optimistic won't be told that their idea won't work, or listen to friendly suggestions that, having given it their best shot, it's time to let it go and try something else. Some economists have gone so far as to suggest that extreme optimists should not be encouraged, or even allowed, to start businesses.

That, of course, is why companies have boards of directors, professional advisors, and regulators, but ample real-world evidence shows that these parties can be persuaded to share the optimistic entrepreneur's "inside view" rather than taking the supposedly detached "outside view." Witness the likes of Elizabeth Holmes, whose Steve Jobs cosplay, complete with black turtlenecks and an artificially deepened voice, persuaded a celestial list of powerbrokers to back her health-tech start-up, Theranos. Holmes appears to have believed her own hype, perhaps thinking she could buy her way to the technology needed to revolutionise blood testing. But she didn't so much "fake it till she made it" as just fake it. Her optimism turned out to be toxic for all involved—herself, her investors and backers, and most of all the patients whose blood tests her company botched.

One might also ask why the founders of some companies have been allowed to retain control over them even after they've become public companies, one of the most notorious examples being Mark Zuckerberg's "golden shares," which make him not so much Meta's

CEO as its supreme ruler; he is, as it happens, obsessed with the Roman emperor Augustus. "Basically, through a really harsh approach, he established 200 years of world peace," he told *The New Yorker* in 2018. Classically attuned observers noted that this was rather euphemistic: Augustus, an autocratic despot, achieved imperial harmony by killing quite a lot of people.

Some might also observe that Zuckerberg's products have also got quite a lot of people killed—from refugee Rohingyas to suicidal teenagers. That highlights another downside of hyperoptimist founders: what makes their product successful can also make it dangerous. For most of its existence, Zuckerberg's Facebook ran according to two principles. One was a commitment to radical transparency: Zuckerberg appears to have genuinely believed that *more* communication could only mean *better* communication; things don't seem to have worked out that way. The other was captured in the famous slogan "Move fast and break things." And again, Facebook's capacity to break things has gone far beyond what anyone might have "realistically" expected.

But with Zuckerberg in charge, there is little that anyone can do about it. Since everyone is on Facebook already (or at least everyone who wants to be), its market position is almost unassailable; but if competitors do look threatening, the company's vast wealth means it can snap them up before they eat into its position, just as it did with Instagram and WhatsApp. And no matter what strategic blunders he makes—as he did by turning Facebook into Meta, with its focus on a metaverse no one wanted—he can't be forced from his throne. Individual optimists gather the resources needed to flourish: skill, money and status. And friends. Well, Meta has plenty of those.

But not for ever. Despite its apparently untouchable status, Meta (and the other Valley giants) is still vulnerable to social and technological change. Kids don't use Facebook: it's for their parents or, increasingly, grandparents. A new approach to antitrust could stop it from devouring promising technologies and companies; some of them might turn out to have a technological edge. Before Facebook, there was Myspace; before Myspace, AOL; before Google there was Microsoft, and before Microsoft, IBM. The average life of a company in the Standard & Poor's 500 index is only fifteen years; veterans get to maybe half a century. Facebook was founded in 2004. It may yet fail to reach a ripe old age.

Philosophers

Once the Valley titans had accumulated more money than they could possibly spend, they moved on to giving it away. And being more inclined to crank out a spreadsheet than show up at the Met Gala, they wanted to do it effectively–which is to say, with the maximum bang for their megabucks. That led to the emergence of "effective altruism," a movement dedicated to ensuring that their donations did as much actual good as possible, rather than satisfying the personal preferences of their founders or other subjective criteria. For example, buying anti-malarial bed-nets saved far more lives per dollar spent than, say, safer drinking water. So the EAs bought tens of thousands of them.

So far, so good. The EA movement centred on the philosophy department at the University of Oxford, where previously obscure academic debates about the morality of prioritising some lives over others were suddenly transformed into real-world experiments backed by billions of dotcom dollars. (When I went to an EA retreat in 2022, at a five-star country-house hotel, I was struck by how closely it still resembled a student debating society, only with vastly more expensive vegan sandwiches.) EA did have its critics–some argued it was patronising to the recipients of its largesse, others that it served as the acceptable face of tax avoidance–but its approach was eminently Leibnizian: considering all the possible options and finding the optimal one.

But things started to go off the rails after that. Members of the EA movement began to make recommendations that were *so* counter-intuitive as to raise eyebrows: for example, that acolytes should become well-paid financiers who could then bankroll further charity, rather than, say, doctors or other front-line professionals. One such financier was Sam Bankman-Fried, whose own hyperoptimistic approach to running a cryptocurrency exchange involved skipping the niceties of running a multibillion-dollar company and landed him in jail.

Bankman-Fried's association with EA reflected poor judgment on the part of some of its founders, but didn't inherently discredit its approach. But then that began to shift, too. EA superstar Will MacAskill contemplated the vast numbers of humans who would never exist if humanity became extinct now—not just those yet to be born on Earth, but in the course of our inevitable colonisation of outer space, and our immortal duplicates uploaded to live eternal lives within superintelligent machines: 10^{38} or even 10^{58} lives. His radically utilitarian logic was somewhat opposite to Schopenhauer's: *any* life was worth living,* so the total amount of human well-being at stake was (literally) astronomical.

Such numbers vastly outweigh absolutely any consideration of the 8 billion people alive today. That meant that averting risks that could wreck human civilisation, or even render us extinct—bioweapons, pandemics, asteroid strikes, nuclear weapons and out-of-control artificial intelligence—was the *real* priority, ahead of even climate change or global poverty. (We'll come back to *how* we might address these risks in Chapter 9.)

* This is questionable: it leads to the so-called "Repugnant Conclusion" that it's better to have lots of people living in abject misery than comparatively few living in comfort.

Extraordinary though this conclusion is, it received considerable attention from many people who should really have known better (aided by a PR campaign of extraordinary magnitude). There are many reasons to question the "long-termist" logic. Some are technical: the ways EAs have estimated extreme risks are unrealistic, and ignore some less flamboyant hazards that are much more cumulatively important. (In short: better error management needed.) Some question implementation: How would you keep up such a programme over vast stretches of time and space? Wouldn't it be better to expend your energies on constructing a society that renews itself every generation, so it sustains itself for ever? And some are pragmatic: if we were to spend *all* our time and energy on securing the well-being of future generations, today's society would fall apart.

All these objections can be reduced to one simple point. The long-termist vision is usually received with a mixture of incredulity and awe. It seems jaw-droppingly, well, visionary. But in fact it's nowhere near imaginative *enough*. It makes the same mistake as Paul Ehrlich of holding some variables constant while extrapolating one—population!—to the stars and back. It takes no account of a future in which humanity decides to stay on Earth rather than fill the cosmos; or where the population ends up at a few million cossetted souls rather than trillions of serfs; or where humanity becomes a single cybernetic super-organism, or whatever else you can dream up.

There's not really any need to evaluate these scenarios in detail, because no one has any idea how likely they are to happen. That's why we don't normally strive to secure the well-being of very distant descendants: instead, we try to look after the planet for the benefit of the next generation or two, or seven, and then hand over to them. But if you *do* try to construct a model, as David Thorstad of Oxford's Global Priorities Institute has, you find that some of these scenarios quickly outweigh the default long-termist scenario, even using its own terms of reference.

All this would be irrelevant—darkly funny, even—were it not for the good that *could* have been done with those billions of dollars. Effective altruism started out as a promising combination of the Leibnizian commitment to optimisation and the Voltairean commitment to gardening: small acts of good, multiplied many times over. That it has ended up pursuing the grandest of all grand plans is, alas, a testament to the hollow allure of Panglossianism.

8

Making Progress

Are things getting better, worse—or neither?

An optimist believes that humans can harness or overcome natural processes in order to make progress. But what does that mean?

Here's an example.

I started my career in a nondescript office block at the top end of Marylebone Lane in London, working for *Risk*, the financial magazine where I learned about derivatives. Marylebone Lane is a strange, off-putting road: it winds sinuously across the West End's otherwise orderly grid of streets, without the wall-to-wall bistros and boutiques of its neighbours. One day, a colleague explained why: the road follows the route of the Tyburn, one of London's storied "lost rivers": it had long since been covered up and built over. For a moment we mused about the Venetian wonderland we might have inhabited had the river never been culverted. The dingy boozer across the road transformed into a riverside *bacaro*; the trudge to the station replaced by a relaxing bankside promenade; black cabs swapped for punts drifting languidly towards the Thames. Our riparian reverie was rudely interrupted by someone demanding to know where on earth their copy was, but I never forgot about the river burbling far below my feet.

But why was the Tyburn buried in the first place? For hundreds if not thousands of years, Londoners had been both drawing fresh water from the Thames's dozen-odd tributaries and dumping waste into them—not just sewage, but food scraps, agricultural run-off and tannery chemicals, animal corpses, industrial waste and household junk. But mostly sewage. The precise history was as murky as their waters became, but the upshot was that over time the lesser rivers, including the Tyburn, turned into noisome nuisances. One by one they were covered over, to carry their cargo unseen into the Thames.

But as the population grew, and more people gained access to running water and flushing toilets, the quantity of waste became so huge that the Thames itself became foul: its waters darkened, filth piled up on the river's banks and the stench mounted. "Through the heart of the town a deadly sewer ebbed and flowed, in the place of a fine fresh river," Charles Dickens wrote in *Little Dorrit*. The scientist Michael Faraday campaigned for action to be taken, writing to *The Times* in 1855, "The river which flows for so many miles through London ought not to be allowed to become a fermenting sewer . . . If we neglect this subject, we cannot expect to do so with impunity; nor ought we to be surprised if, ere many years are over, a season give us sad proof of the folly of our carelessness."

In fact, the subject had already *been* neglected for some years: an ambitious plan to revamp London's totally inadequate sewerage system had been proposed but summarily rejected as unaffordable. And so the season Faraday predicted came three years later, when a heatwave led to London being suffused by the "Great Stink." It was unendurable in the House of Commons, where curtains soaked in "chloride of lime" failed to alleviate the stench; by one of those remarkable coincidences so common in politics, it was only then that parliamentarians agreed something must be done—although one Tory continued to argue that the government "had nothing whatever to do with the state of the Thames." *Plus ça change*.

Civil engineer Joseph Bazalgette promptly revived his plan for a system of properly engineered sewers that would carry waste out of the city. More than a thousand miles of drains ran into the newly brick-lined rivers Tyburn, Fleet and Walbrook, now pressed into service as "interceptors" which ran down to the banks of the Thames. They didn't empty into the bigger river, but into two further tunnels which ran alongside it to the estuary, where pumping stations ensured the waste was washed out to sea with the tide.

Bazalgette thought big, and he thought for the long-term. The project required 318 million bricks, 670,000 cubic metres of concrete and thousands of labourers working for nearly twenty years; it also used innovative materials, notably highly water-resistant Portland cement. Budgets were blown and deadlines missed—the planning fallacy was as potent then as it is now—but there was only one chance to get it right. I've visited one of the sewers he built, and also one of the pumping stations: both were beautifully constructed marvels of engineering (albeit that you did, of course, have to look past their contents to appreciate that).

And they worked. The project led to a far cleaner, greener city, and a much healthier one. Londoners had been dying in their tens of thousands from cholera; the new sewage system put a stop to the transmission of that and other diseases, with enormous benefits for Londoners' health. It also had other benefits: twenty-two acres of land were reclaimed, creating the riverside embankments which still define the city today, and under which the District and Circle Underground lines carry hundreds of millions of passengers each year.

Bazalgette's achievement has become the stuff of engineering legend. Consider, for a moment, the scale of the challenge. To early Londoners, the Thames must have seemed inexhaustibly vast. Treating it as an open sewer might not have been especially savoury, but there seemed little chance of exhausting its capacity. How could it possibly become so polluted as to be utterly toxic?

And then, once it became apparent that it had, how could mere humans possibly clean it all up again? It must have seemed—well, it must have seemed as formidable as the challenge of purging the sky of greenhouse gases does today. We'll come back to that.

The New Optimists

Progress, for Hans Rosling, comprised good things on the rise. "People often call me an optimist, because I show them the enormous progress they didn't know about. That makes me angry. I'm not an optimist. That makes me sound naive," he wrote in *Factfulness*, a book which has become something of a bible to development nerds since its publication in 2018. "I'm a very serious 'possibilist.' That's something I made up. It means someone who neither hopes without reason, nor fears without reason . . . I see all this progress, and it fills me with conviction and hope that further progress is possible. This is not optimistic. It is having a clear and reasonable idea about how things are."

Rosling, a professor of international health, might not have liked being called an optimist, but he certainly seemed like one. He'd joined the sparse ranks of superstar statisticians—alongside election-night analysts and football pundits—by giving energetic, upbeat talks which demonstrated that the world was not in such terrible shape as you might think. To make his case he drew extensively on data presenting positive trends in everything from birth rates to bear attacks, arguing compellingly that today is in many respects the best possible time to be alive.

Rosling's informal approach and penchant for theatrics—he closed the second of his many TED Talks by swallowing a sword—won him celebrity with both the YouTube audience and the Davos elite. But his arguments didn't always persuade more buttoned-up academics. As one critic put it, his presentations covered "bad things in decline" and "good things on the rise," but never "bad things on the rise." Rosling did have a point, however, when he told his audiences, whether students or billionaires, that they had failed to recognise the reality of progress in many countries—obscured as it was by outdated stereotypes and the disproportionate focus on grim news stories.

He was not alone. Steven Pinker suggested in his 2011 book, *The Better Angels of Our Nature*, that the evidence shows the world to be more peaceful than it has ever been, thanks to the combination of democracy, trade and globalisation. He doubled down in 2018's *Enlightenment Now*, having subsequently realised that there had also been dramatic improvements "in prosperity, in education, in health, in longevity, in child mortality, in attitudes towards women, ethnic minorities and gay people, even time spent on housework."

Others joined the chorus: Matt Ridley; the Swedish historian Johann Norberg, author of *Progress*; Max Roser, founder of Our World in Data, something of a successor to Rosling; and various journalists and academics who chipped in from time to time. Many, but not all, opposed those sceptical of the socioeconomic status quo, or discontented with the populist turns of 2016. While never an organised group, they were collectively dubbed the "New Optimists" by the writer Oliver Burkeman, because of their shared commitment to the idea that dispassionate, data-driven analysis demonstrates that the global population is doing much better than at any time in the past, with positive long-term trends obscured by the media's fixation on evanescent bad news and cries of alarm from gloom-peddling commentators. In some cases, they were making the rather Whiggish case that we have found a winning formula: Pinker, for example, championed the "Enlightenment values" of reason, science, humanism and progress.

These claims did not go unchallenged. While many readers applauded Pinker as painting a cool-headed, clear-eyed picture of the state of the world, others accused him of wearing rose-tinted glasses. Critiques were made of his choice of data and methodologies, his account of the Enlightenment and its thinkers was pooh-poohed, and Nicholas Taleb squabbled with him over the proper use of statistics. Others suggested that the analysis simply wasn't meaningful: How does the existential risk of nuclear annihilation compare to the more visceral threat of being bludgeoned to death by a raiding party? Echoes here, once again, of the Leibniz–Voltaire contretemps: How can *this* possibly be the best of all possible worlds?

The New Optimists' heyday came just after the democratic upsets of 2016, but their arguments have not so much gone away, despite continuing turbulence in global affairs, as they have been assimilated into the ongoing dialogue over the reality and nature of progress. December round-ups of all the wonderful, untrumpeted advances and breakthroughs that took place over the course of the year have become something of a tradition for some pundits, usually prefaced with a defiant or apologetic acknowledgment that they are swimming against the tide. (I plead guilty to having penned a couple of these myself.)

Some elements of the New Optimist manifesto are difficult to argue with. Many key indicators of human thriving have, indeed, improved dramatically over the course of the past two centuries, particularly in the second half of the twentieth century. It's also broadly correct to assert that "the media" is biased towards negative and attention-grabbing stories, while steady, continuous improvements in the human condition go unreported. As Roser tweeted in 2017, "Newspapers could have had the headline 'Number of people in extreme poverty fell by 137,000 since yesterday' every day in the last 25 years."

It is well worth reading Rosling and Pinker for a reminder of just how far we've come, and how quickly we've got there. It's very much worth exploring Our World in Data to reality-check your assumptions about the state of the world. But what lessons do they hold for the future? The obvious conclusion is that we should gain conviction in what we've been doing and hope it will continue to work.

But should we? There are three reasons we should be cautious.

The first is that it's hard for data to reflect, or capture, how people *feel*. Most of us have to take the New Optimists' arguments on faith, since we have neither the data nor the skills to check them—or, for that matter, the counterclaims of their critics. Sites like Our World in Data do allow you to look at the figures for yourself, but interpreting them is not straightforward, as noisy rows over issues like the meaningful definition of poverty demonstrate. And data is by its nature impersonal. It's long been the case that people's fear of crime has little to do with their actual risk of being involved. The disparity is probably due to the availability heuristic: crime tends to hit the headlines, and therefore to stand out in our minds.* Sure, maybe violent crime is down over the past 300 years, and that's reassuring in an abstract kind of way, but what I really care about is what I hear has happened in my neighbourhood. Now a similar discrepancy is starting to show up in economics: people feel much worse off than the data say they are. And how they feel matters, both in itself and because of how it informs their actions.

The second is that New Optimism lends itself to complacency. Yes, we have made tremendous progress on a variety of fronts; but perhaps they could be even better than they are. We have no assurance that this is the best of all possible worlds: there are certainly plenty of ills that still assail us. And there are no guarantees that the way we've gone about things in the past will continue to work in the future—particularly when it comes to resource consumption, or environmental limits. A good chunk of the improvement in our material standards of living has been achieved by pillaging the planet in a way that may reverse those improvements for generations to come: that's certainly what some in those generations believe. The party may be over; the morning after has yet to come.

* I believe I introduced Pinker to the newsroom maxim "If it bleeds, it leads" back in 2011. Call it my contribution to the New Optimist manifesto.

The third is the potential for massive technological change—whether positive or negative, or, as is often the case, a mixture of both. Nuclear weapons have only been used twice, but have shaped the geopolitical trajectory of the planet since 1945; electronic computers, born of the same conflict, allowed us to solve problems—like weather forecasting—that had previously been utterly intractable. The most obvious candidate in the twenty-first century is artificial intelligence, but there are others: nuclear fusion could offer essentially limitless energy; biotechnology could reduce or even remove our dependency on the Earth's biosphere; geoengineering could radically change the very sky and sea. Under those conditions, the very concept of progress could change for ever.

The Pursuit of Happiness

Try to visualise progress, and there's a good chance you will picture a line, or arrow, going up and to the right. But what does that line represent?

Your first impulse might well be to say that it's gross domestic product (GDP). Certainly, that's all that many politicians and powerbrokers seem to have cared about for decades. GDP measures the size of a country's economy, so changes in GDP over time indicate the growth, or shrinkage, of that economy. In some countries, increasing GDP—which is to say, boosting economic growth—has become the beginning and end of all political and social debate. Get growth right, and all else will follow: not just financial well-being, but good health, clean environments, strong security and just plain well-being.

Can it sustain such hefty expectations? It's not what it was designed for. In the 1930s the American economist and statistician Simon Kuznets devised a way of calculating gross national product (GNP), the total value of goods and services produced by a country's residents, in order to provide some much-needed clarity about economic activity during the Great Depression. Prior to Kuznets's work, there had been no such measure, so GNP was a genuinely valuable tool for a government trying to steer its national economy out of the mire—all the more so during the Second World War and its aftermath, when governments were forced to take direct control over their countries' economies. This is also when resource consumption began to skyrocket, propelled by the war, subsequent reconstruction and the emergence of the consumer economy, a process historians call the "Great Acceleration." GNP, too, increased rapidly, and became enshrined in many places as a single number by which progress could be measured, although as Kuznets told Congress in 1934, "The welfare of a nation can scarcely be inferred from a measurement of national income."

It wasn't that its flaws went unrecognised. In March 1968, Robert Kennedy, launching his campaign to become the Democratic presidential candidate, told 20,000 students at the University of Kansas that "Gross National Product counts air pollution and cigarette advertising . . . It counts the destruction of the redwood and the loss of our natural wonder in chaotic sprawl . . . It counts napalm and counts nuclear warheads. Yet [it] does not allow for the health of our children, the quality of their education or the joy of their play. It does not include the beauty of our poetry or the strength of our marriages, the intelligence of our public debate or the integrity of our public officials . . . it measures everything, in short, except that which makes life worthwhile."

Whether Kennedy would actually have moved away from GNP if he had completed his run at the presidency is a question for counterfactual history.* As it was, he was assassinated three months

* Several have been written, including one by RFK's former speechwriter Jeff Greenfield, in which President Robert F. Kennedy is an Obama-esque figure who "presides over a country in which war and partisanship have wrought growing disaffection," according to Michiko Kakutani in the *New York Times*.

later, Richard Nixon took the White House, and GNP continued to be a key metric of success for politicians. As the world globalised, GDP—the total value of goods and services produced within a country's boundaries—became the preferred measure, enshrined not only for industrialised countries, but for others aspiring to be noticed on the international stage as well. A country could not afford to slip behind on GDP, lest international borrowing arrangements be jeopardised or investor confidence shaken. All policy was subsumed into economic policy, until by 1992, Bill Clinton could simply declare: "It's the economy, stupid!"

But RFK was right. Three decades of politicians managing GDP, increasingly to the exclusion of everything else, has made all the things it doesn't capture all the more apparent. GDP counts the amount of money in the economy but pays no heed to who has it: so wage or wealth inequalities are irrelevant. It pays no attention to carbon emissions or other waste products at all, so environmental degradation doesn't show up. And a growing proportion of the global economy, particularly in WEIRD societies, is based on services and information technology, neither captured well by standard GDP calculations. In short: GDP might've been a radical innovation for the twentieth century, but it is ill-suited to the challenges of the twenty-first. It's not *just* the economy. Stupid.

So what should we target instead? One increasingly popular answer stems, unexpectedly, from the mind of a teenaged king of Bhutan. Having taken the throne of this tiny Himalayan kingdom in 1972, at just sixteen years of age, His Majesty the Fourth Druk Gyalpo Jigme Singye Wangchuck, declared that "gross national happiness" (GNH) should take precedence over "gross domestic product." Ostensibly this was in response to what he'd heard from his new subjects, but it also could have been a retort to unfavourable commentary on his country's extreme poverty, as gauged by more conventional metrics. It sounds like the sort of thing many of us might've come up with as idealistic teenagers, and in fact GNH was little more than a quirk of Bhutan's reputation for the next couple of decades. In practice, it was used to rebuff calls for reform and reinforce the dominance of the ruling elite. You may be poor, but it says here you're happy.

But the idea had more substance than its origins suggested. It's been known for decades that happiness and the health of the economy aren't necessarily related. In 1974, the economist Richard Easterlin looked at the relationship between happiness and income (per capita GDP) in the US between 1946 and 1970, and found that while people with higher incomes are happier than poorer people at any given time, they don't get any happier if there's a general increase in income. Why? Perhaps because as social primates we like to *feel* better off than our neighbours—thus validating our comparative optimism; if we all get richer, that gap stays the same. And if the riches are spread unevenly, the 1 per cent might get happier but the working stiffs more miserable.

The Easterlin Paradox, as it was known, was largely dismissed as a curiosity rather than a serious consideration. But in 2008, with a new king on the throne and democratic reforms underway, Bhutan wrote GNH into its new constitution, charging its government with maximising psychological well-being, health, time use, education, cultural diversity and resilience, good governance, community vitality, ecological diversity and resilience and living standards. Three years later, Bhutan sponsored a UN resolution calling on international governments to "give more importance to happiness and well-being in determining how to achieve and measure social and economic development."

Leaders elsewhere were ready to give the idea a hearing. In November 2010, the UK prime minister, David Cameron, announced that it was "high time we admitted that, taken on its own, GDP is an incomplete way of measuring a country's progress," and asked the Office of National Statistics to compile figures on national well-being based on 58 measures in 10 categories, including health, economics, government and environment. It duly did so, but other than periodic clickbait about the country's "happiest towns" there has been little evidence of these efforts since—perhaps because the announcement was swiftly followed by a protracted period of austerity and the divisive Brexit referendum, neither of which greatly enlivened the citizenry. Happiness is harder to fudge than economic measurements.

The story has been similar elsewhere. While a host of countries and municipalities signed up to the happiness agenda, many of their announcements turned out to be little more than short-lived feel-good froth. One notable exception was New Zealand, which in 2019 issued a "well-being budget" sidelining GDP in favour of spending that addressed mental health, child poverty, inequalities affecting indigenous peoples, digitalisation and the transition to a sustainable economy. While widely applauded, there was scepticism that the same approach could work outside the context of a small, insular nation like New Zealand—or Bhutan, whose self-imposed isolation has made it hard for outsiders to gauge how successful its happiness drive has really been. It remains both poor and not obviously happy, with many of its young people emigrating in search of a prosperous future elsewhere.

Nonetheless, the happiness agenda still has momentum behind it, with Nobel laureate Joseph Stiglitz at its forefront. Stiglitz initially led a commission on the measurement of economic performance and social progress for French president Nicolas Sarkozy, which rolled into a similarly themed working group at the Organisation for Economic Cooperation and Development. That group reported in 2018 that countries would be better served by adopting a dashboard of economic, health and social indicators, which the OECD duly published as an 11-indicator "Better Life Index." In the 2020 edition, covering forty-one countries, Norway came out on top; South Africa at the bottom. Life had got better since the 2010 edition, the OECD reported cautiously, but "recent advances in well-being have not been matched by improvements in the resources needed to sustain well-being over time. From financial insecurity in households, through to climate change, biodiversity loss and threats to how democratic institutions perform their functions, we need to look beyond maximising well-being today. Ensuring continued prosperity for people and the planet will require bold and strategic investments in the resources that underpin well-being in the longer run." Or to put it another way, perhaps the New Optimism is about to run out of steam.

Deciding on those investments is not easy. Different countries face different challenges, and resolving them isn't necessarily straightforward: there are compromises and trade-offs needed—and those, Stiglitz and his colleagues suggested, require each country to engage in "robust democratic dialogue to discover what issues its citizens most care about." We're back to Leibniz's compossibility: you can fix one problem in isolation, but if you fix each of your problems one by one, you won't necessarily improve the overall state of affairs, particularly if you have a limited budget to work with.

For decades, we've been acting as though GDP is the determinant of optimality; but that confidence seems to have been either misplaced, or no longer appropriate. Maybe, despite Kuznets's reservations, it once reflected well-being, when living standards tracked the explosive growth of the Great Acceleration, but as the adage goes, "When a measure becomes a target, it ceases to be a good measure." That, it seems, happened years ago.

The World Is Awful

The natural successor to Hans Rosling is Max Roser, an Oxford-based economist who collects data about human thriving, including the kinds of basic metrics that Rosling highlighted, those used to calculate Gross National Happiness and the OECD's Better Life

Index, and many more. These metrics are published as interactive charts on Roser's website, Our World in Data, and are often turned into snappy infographics circulated on social media by the likes of Bill Gates, whose foundation funds Roser's work. Roser, like Rosling, says he's motivated by a desire to demonstrate that it is, in fact, possible to improve the lives of those in poverty, even if there's still a long way to go: "The world is much better; the world is awful; the world can be much better," as he puts it, in true meliorist style.

But Roser, like Rosling, also has his critics, who claim that his charts support particular, misleading narratives: those neat up-and-to-the-right lines hide more complicated truths about the nature of progress. "Prior to colonisation, most people lived in subsistence economies where they enjoyed access to abundant commons—land, water, forests, livestock and robust systems of sharing and reciprocity," the anthropologist Jason Hickel wrote in 2019. "They had little if any money, but then they didn't need it in order to live well—so it makes little sense to claim that they were poor." In Hickel's telling, money only became relevant once they'd been abstracted from their previous lives, into formal employment and consumerist lifestyles. Claiming increased earnings meant better lives was to ignore the bigger truth that their lives might have been better before they had any earnings to speak of at all.

There's plenty of reason to doubt Hickel's account of idyllic precolonial existence, given that people have fought, starved, suffered and pillaged since time immemorial. But it is also true that abstracting people from their traditional lifestyles, as the march of "progress" often does, can represent a loss of well-being that isn't reflected in economic statistics. Melting ice is opening up vast mineral wealth in the Arctic; but many Inuit are sceptical that the consequent boost to GDP will improve their lives or make up for their solastalgia.

Hickel's opposition to conventional measures of growth makes sense given that he's a leading voice in the degrowth movement.

Degrowth is a broad church, but its advocates generally agree that the planet can't tolerate much more exploitation in the name of growth; and since GDP doesn't adequately capture well-being, that pursuit is misguided anyway. At its soft end, that simply means de-emphasising growth in decision-making; at the hardline end, it can involve economic ideas that sound outlandish to orthodox economists—just as some of today's orthodox economics once sounded outlandish to a previous generation.

Modern monetary theory (MMT), for example, suggests that governments today shouldn't worry about borrowing as much as they need, since they can always print as much money as is required to pay the interest. The proceeds of the debt can then be used to pay for giant programmes like universal healthcare, climate adaptation and employment guarantees. MMT became popular because it seemed to fit with the economic environment after the financial crisis; it was how Alexandria Ocasio-Cortez proposed to pay for a Green New Deal—a package of measures intended to tackle not just climate change and other environmental woes, but a host of other issues, such as job creation and economic inequality. Like Franklin Roosevelt's original New Deal, it aimed at comprehensive reform rather than incremental improvement—in a sense, the realisation of an entire better world. But it failed to pass the Senate; one reason was its reliance on an "unproven" economic approach.

Mainstream economists may have little time for MMT, but they increasingly agree that we do need to rethink our approach. "Adam Smith wrote *The Wealth of Nations*, not the GDP of Nations," says the distinguished environmental economist Partha Dasgupta, who expresses no doubt that his profession's growth-dominated framing of each and every issue should give way to something more holistic and reflective of *all* the wealth of nations, including their natural assets. And perhaps in the long-term we can be optimistic about the potential to move away from the dominance of extractive growth towards . . . something else. Degrowth, circular economies, doughnut economics, resilience: so far no name has stuck. The problem, as ever, is that the "something else" is unclear and untested: it looks inherently unrealistic compared to the status quo. But is it really?

The notion of degrowth could stiffen our resolve in areas that have so far proven resistant to conventional policymaking. A well-designed carbon tax, for example, could go a long way towards suppressing highly emissive consumption in rich countries. So far, this has proven politically unpalatable and socially infeasible. But attitudes are changing: there are indications that Europeans, at least, are now willing to prioritise the environment over growth. And policymakers are also signalling intent: the IPCC's 2022 report for the first time suggested degrowth as one way of mitigating the effects of climate change.

There are plenty of indicators suggesting that consumption is already "decoupling" from environmental damage—in particular, the increasing divergence between carbon emissions and GDP in wealthy countries like the UK. The astonishing progress made in reducing the cost and increasing the practicality of renewable energy means that, in many instances, it's now simply more competitive than fossil fuels. The implication is that eschewing growth for the sake of it would remove the incentives that are driving that transition; it could also whisk away the ladder from those still in poverty.

But neither degrowth nor decoupling can be relied upon. Decoupling may not get us there fast enough, while degrowth is anathema to the economic and political establishment. And the two sides both have a tendency to dismiss each other as lunatics intent on wrecking our well-being. The degrowthers claim the decouplers are Pollyannas playing the Glad Game while the planet burns; the decouplers say the degrowthers are innumerate lunatics who will wreck the progress we've made. Roser and Hickel are adversarial to the point of animosity. Watching from the sidelines, one has the sense that ideology is winning out over pragmatism: posturing arguments about the perfect solution are obstructing the search for the best possible one. Compromise, much less discussion, doesn't seem to be on the table.

As a 2022 *Nature* editorial marking the fiftieth anniversary of *The Limits to Growth* put it: "Research can be territorial—new communities emerge sometimes because of disagreements in fields. But green-growth and post-growth scientists need to see the bigger picture. Right now, both are articulating different visions to policymakers, and there's a risk this will delay action. In 1972, there was still time to debate, and less urgency to act. Now, the world is running out of time."

The End of Ideas

Given the conceptual and practical problems of economic growth, some people have started to put their faith in another force to carry the day: technology. Since its very beginnings, science fiction has written breathlessly of how technology can cure all ills. But techno-optimism became a force to be reckoned with when the internet became widespread in the late 1990s—offering seemingly miraculous new capabilities.

The internet, it was said, would revolutionise commerce, provide limitless free entertainment, generate unprecedented opportunity for creators and coders alike, educate the world, spread democracy, destroy all those old boring companies everyone hated (like the record labels) and incidentally make a few people on the Sand Hill Road absolute gobs of money. And in fact, it would do this regardless of whatever anyone had to say about it. Disruption was the order of the day: moving fast and breaking things, as Zuckerberg put it. That mutated into a belief that it was technology *in itself* that did these things, if only it was left unmolested by regulation, ethics or other constraints on its power.

In 1999, the inventor Ray Kurzweil published *The Age of Spiritual Machines*, in which he elevates Moore's Law—the observation that the number of transistors on an integrated circuit doubles every two years—into, essentially, a fundamental law of the universe which suggests that the power of technology will continue to accelerate until it ultimately surpasses and encompasses humanity, ushering in utopia. But despite its name, Moore's Law is not any kind of law; it's a fine example of how optimism can be self-fulfilling. People think it's true, so they find ways to *make* it true; the underlying technology has changed several times since Moore coined it. It is not a law of the universe: it is, once again, simply drawing the line up and to the right.

In 2010, the former *Wired* editor Kevin Kelly—a self-declared optimist—wrote *What Technology Wants*, a description of the "technium," the all-encompassing system of technology that surrounds us. This, he was keen to explain, had evolved like a biological system, and would continue to do so—weirdly divorced, in his account, from the humans who actually have to invent, design and build it. And that meant it would in time deal with problems like overpopulation and resource depletion, when it got around to it, "just, I suppose, as biological evolution helped the dinosaurs deal with that meteorite," wrote the biologist Jerry Coyne in the *New York Times*.

And in 2023, the venture capitalist Marc Andreessen published a "Techno-Optimist Manifesto" in which he drew upon the idea of "effective accelerationism"—a hybrid of effective altruism and accelerationism, a school of thought which derives from a small but influential cadre of esoteric thinkers based at the Cybernetic Culture Research Unit at the University of East Anglia in the late nineties. Accelerationism suggests that technology should be encouraged to develop as fast and freely as possible: take any and all brakes off innovation, it suggests, and everything else will follow: growth, prosperity, well-being and happiness. Written in snippets, like the world's longest Twitter thread, Andreessen's manifesto was a nakedly self-serving document, calling for freedom from regulation—and any other constraint on, say, a venture capitalist's ability to make money—in the name of progress.

Of course, technology *does* fuel progress. It is the fundamental reason that our lives today are so much better, by almost any measure, than they would have been five hundred years ago, or even fifty years ago. The Great Acceleration was born of rapid improvements in technology. But *pace* Kurzweil, Kelly, Andreessen and a host of others, it takes humans to achieve that. As the economist Marianna Mazzucato has convincingly demonstrated, the seeds of Silicon Valley's success were planted and cultivated decades earlier—in government-funded labs. That's even truer when it comes to inventions that involve actual stuff, rather than information: think of Katalin Karikó and her mRNA vaccines.

The problem is that there seems to be less of that breakthrough innovation coming through the pipeline. An influential study by a team of researchers from Stanford and MIT concluded in 2020 that ideas, and the exponential growth they imply, are getting harder to find: to pick one headline statistic, it takes eighteen times as many researchers to keep Moore's Law ticking over today as it did in the first few years after its formulation. The reasons why are unclear: it might be that all the good ideas have been taken, or it might be that scientific research has got too cumbersome and expensive. Or a combination of both: we're seeking big wins—with grand experiments like sending people *back* to the Moon—while more modest projects miss out on funding and attention.

That's spawned a new field: progress studies, founded by the economist Tyler Cowen and the billionaire entrepreneur Patrick Collison. This isn't wholly new territory—the study of science and technology has been established for many years—but Cowen and Collison intended theirs to kickstart progress, whatever form it takes in the twenty-first century, rather than merely observe and comment on it: "[Progress studies] is closer to medicine than biology: the goal is to treat, not merely to understand," they wrote in their initial call to arms, published in *The Atlantic*. What they are aiming to "treat" is GDP per capita—from which they believe all else will follow. Whether that's what technology wants, or humanity needs, remains to be seen.

The Whitening Sky

Let's go back to where we began this chapter, with the Great Stink, and what it can teach us about cleaning up the sky. The parallels are striking. Both then and now, the combination of scientific denial and political intransigence has delayed progress. Both then and now, the problem has been caused by too much of a good thing—the Industrial Revolution then, the Great Acceleration now—that has primarily enriched the WEIRD world while the poor have ended up stuck with the tab. And, like the Victorians, we're daunted by the scale of the challenge. We're not sure that we can build big enough and quickly enough. We know the work will have to continue long after most of us alive today are dead. And we're divided about whether the job requires government action or can be left to commercial forces.

But history doesn't repeat, it rhymes. Nineteenth-century London was enriched by its vast empire, and could act unilaterally and autonomously to sort out the Thames. Climate change is a global problem, and will require a more collective approach to solve. But the point isn't that we should mimic the solutions of an earlier age: it's that we should be inspired by them. Their stories can encourage us to believe—as unrealistic as such expectations may at first seem—that our own problems aren't as insurmountable as we think. The Victorians did it. Perhaps we can, too.

The climate-change equivalent of this Victorian megaproject is geoengineering, the use of technologies that would actively manage the Earth's climate on very large scales, up to and including interventions that would affect the entire planet. It comes in two main varieties: either removing greenhouse gases from the atmosphere, or preventing sunlight from reaching the Earth's surface. There are a number of approaches to both, requiring varying degrees of technological sophistication.

Greenhouse gas removal, for example, can be achieved by planting trees or crops, then harvesting them while capturing and storing the carbon they've taken out of the atmosphere. The same effect could be achieved by seeding the ocean to encourage algal growth. There are also inorganic options, such as boosting the capacity of rocks to absorb carbon by crushing them into gravel or dust, which reacts more quickly with atmospheric carbon dioxide. Or it can be more technological: building massive industrial plants to chemically extract greenhouse gases from the air, which can then be stored or put to industrial uses.

Solar radiation management, similarly, comes in low- and high-tech varieties. It can be effected as simply as painting roofs white and planting light-coloured crops—or covering areas of wilderness with reflective sheeting, perhaps to protect features like significant glaciers. Or it could involve injecting vast quantities of tiny reflective particles—sulphur dioxide is the usual candidate—into the upper atmosphere, to artificially simulate the effect of volcanic eruptions like Mount Tambora. It could even involve launching gargantuan shields or mirrors into orbit to block or reflect sunlight away from the Earth.

Re-engineering the climate of an entire planet requires vast amounts of . . . well, everything. Planting trees is simple, but would require huge tracts of land, much of which has already been claimed for other purposes. Carbon-absorbing infrastructure would take large quantities of material and energy to build—and that in turn means huge costs and, for the moment, huge carbon emissions. Many of the technologies needed have not yet proven themselves at industrial scales, or are not yet economically viable. Some simply don't exist. And because the climate is a family of complex, sometimes chaotic, systems, any solution will inevitably have unintended consequences that are hard to predict and control.

For all these reasons, geoengineering is no one's preferred approach to tackling climate change, just as lockdown is no one's preferred approach to tackling a pandemic. Like lockdowns, the best justification for geoengineering is that it could buy us time to implement a better solution: the decarbonisation of our societies and industries. We've seen already that the climate is incredibly complex: meddling with it is an unappealing and potentially risky proposition. But it's not clear that decarbonisation can or will happen fast enough to avoid deeply harmful levels of climate change either, whatever the decouplers may hope.

We may find ourselves without much option. "We live in a world where deliberately dimming the fucking sun might be less risky than not doing it," Andy Parker, the project director for the Solar Radiation Management Governance Initiative, told Elizabeth Kolbert, author of *Under a White Sky*, an exploration of human efforts to "fix" the environment. The title refers to the changed appearance of the sky under the stratospheric-injection version of geoengineering. A technology with such transformative effects is daunting to contemplate: many climate specialists view it as not just unrealistic but unthinkable. But an increasing number of scientists, if still a minority, disagree; and some wealthy activists are ready to back them. Guerilla projects are already taking place around the world, to the dismay of some and the delight of others.

Let's say we come to the point where the answer to Parker's dilemma is clear-cut: the planet is wracked by extreme weather, our societies thrown into turmoil by massive destitution and migration, our very civilisation threatened by widespread crop failures and the food shortages predicted by Paul Ehrlich. It's clear we need to take drastic action. From the above, it's clear that embarking on a large-scale geoengineering programme would be a profoundly optimistic project. Could we make it work?

First, we'd need close global cooperation. That seems risible, after a succession of climate summits that have resulted in what Greta Thunberg would regard as "blah, blah, blah." But there is a precedent. In 1985, scientists discovered a vast, and growing, hole in the ozone layer of the stratosphere above the Antarctic—ozone being the form of oxygen that protects life from the Sun's harsher, ultraviolet radiation. The culprit had been identified a decade earlier: CFCs, chemicals used in refrigeration. A UN treaty signed *that same year*, the Vienna Convention, committed the international community to further research; two years later, the Montreal Protocol was signed, committing signatories to phasing out CFCs. These remain the only UN treaties ever to have been ratified by every country on Earth.*

Is it likely that we can replicate this success? Ozone was a simpler problem, and there was a "smoking gun"; but so there is with climate, now that its effects have become inarguable—and dangerous. Ozone only affected one industry; with carbon there are legitimate fears of "moral hazard"—that carbon-removal technologies, in particular, could be used by carbon-intensive businesses as a rationale to carry on as usual. But that, too, could be addressed if the political and social will was there. And it does seem to be turning that way: regulation, disinvestment and consumer distaste is moving against such businesses. Susan Solomon, the scientist whose work galvanised the UN ozone treaties, thinks the lessons are transferrable: her third book is called *Solvable: How We Healed the Earth, and How We Can Do It Again*.

How about the technology? The difficulties involved in, say, sucking carbon dioxide out of the air at industrial scales are

* They are still being revised and implemented to this day—and still paying off in unexpected ways. CFCs are also powerful greenhouse gases: outlawing them is estimated to have delayed the transition to an ice-free Arctic by around fifteen years.

formidable. But let's first consider the other side of the equation: reducing the amount of carbon we are putting *into* the atmosphere. The first solar cell, devised all the way back in 1883, converted just 1 per cent of the sunlight falling on it into electricity. That didn't change meaningfully for nearly a century: solar power remained a wildly expensive joke. But the US energy crisis in the seventies, and then mounting awareness of climate change, provided incentive to develop it more rapidly. The rate of acceleration was spectacular. A study by researchers at the University of Oxford in 2021 looked at more than 2,900 estimates of the cost of solar power between 2010 and 2020. The average forecast was a 2.6 per cent annual reduction; the highest was 6 per cent. The actual rate? Fifteen per cent. Decarbonisation of the electricity grid, once a wicked problem, has turned into a self-fulfilling prophecy.

This isn't to say that new technologies will solve the crisis "just like that," in Thunberg's derisive words. Some will ramp up faster than anticipated, some will struggle and some will simply fail. But we won't know which until we start trying to build them in earnest—and that means creating incentives for them to be developed and deployed at speed. The incentives that created the "Green Vortex" of renewable energy weren't those recommended by economic orthodoxy, but subsidies, cheap finance and public procurement. Those might do the trick again in some cases; or we might turn to innovative financing for innovative technology.

For example, there's an "advance market commitment" from a consortium of multinational businesses, led by progress enthusiast Patrick Collison's Stripe, to buy at least $1 billion of carbon removal between 2022 and 2030. The idea is that this will demonstrate to entrepreneurs that there's a market for their technology—whose economics are currently unrealistic, just as solar energy's once was—thus making it easier for them to justify investing money and talent in new ideas. Just such a commitment was previously used to encourage pharma companies to develop an affordable vaccine against pneumococcal illnesses. It worked: once the vaccine existed, governments and philanthropists bought hundreds of millions of doses. It's claimed to have sped up rollout by half a decade and saved around 700,000 lives.

Optimism doesn't appear on any balance sheet, but the investments it encourages certainly do. Similarly optimistic investments in geoengineering technologies might likewise unlock economies of scale sooner rather than later. And we do need to get started sooner. If we do not start work on these technologies now—at full tilt, with all the resources we can bring to bear—we may find that they are not ready when we realise we really need them. Again, there's the medical analogy: we knew what we'd need when a pandemic arose; we should have stocked up when we had the time. (And of course there's the planning fallacy to think about: we may crack the technology, but turning it into infrastructure may well take longer than we hope.)

Right now, however, there's effectively a moratorium on even basic research into the more radical forms of geoengineering. There's an understandable justification for that: fear of unintended consequences. In 1955, the brilliant mathematician John von Neumann wrote in *Fortune* that actions to control the climate "would be more directly and truly worldwide than recent or, presumably, future wars, or than the economy at any time . . . [and] merge each nation's affairs with those of every other, more thoroughly than the threat of a nuclear or any other war may already have done." But with more individuals, companies and states worried about climate and motivated by the trillions of dollars in play, that research may start anyway—unregulated and unfettered. Panglossianism is at work: there is money and power to be had, and the promise of a brighter future. We need to be able to predict the consequences of geoengineering before someone starts doing it: that means conducting experiments and building models.

That would need to take in the social and cultural ramifications, as well as the technological. Seeding the sky with sulphates would turn it a hazy white; seeding the ocean with iron would turn it murky green. While that might not affect human well-being too badly—not as badly as runaway climate change, at any rate—such a dramatic, and clearly artificial, change in our environment would no doubt come as a solastalgic shock of the highest order. We might get used to it. We no longer miss the stars in the night sky, drowned out by light pollution; nor do we really question the aircraft contrails scribbled across it by day. Perhaps in time we'd get used to it being bleached out, too. The art created in the wake of Tambora proved enormously influential; perhaps that inspired by the bleaching of the sky would, too. That, though, is a question for poets, artists, historians, philosophers: the kind of people who gathered in Cove Park that November morning in 2020.

We also need to think for the long-term. Even if the technology works as intended, we're signing up to maintain it for many decades to come, perhaps even centuries. Some forms of geoengineering, like stratospheric sulphate injections, have to be carried out continually; any hiatus while the atmosphere is still high in greenhouse gases would lead to a disastrous crash. Do we want that responsibility? The Victorians, again, provide a precedent. Rather than attempting to turn back the clock, and return the Thames and its tributaries to their long-vanished natural state, they adopted a transformative solution—one that acknowledged that the rivers would simply never be the same again. In effect, they had taken control of them for perpetuity.*

And in fact, I held out on you earlier: their solution is now coming apart at the seams. Bazalgette's sewerage system combined drainage and sewage. Now it's overloaded in both respects. Heavier and more frequent rain is making it overflow—and when it does, the greater load of effluent ends up in the Thames. History is repeating: both the problem, and the solution: London has opted

* Of course, human activity always changes nature, and has done so on larger scales and for longer times than we may appreciate. The medieval London Bridge, for example, dramatically changed the flow of the Thames, making it dangerous for boats to navigate and ultimately reshaping the river basin itself.

for another grand plan: another, much bigger sewer, to supplant Bazalgette's. But to deal with the flash floods triggered by climate change, it will have to garden, too, that is, become a "sponge city" which incorporates areas that catch and absorb water—essentially, green space like parks, woods and wetlands. Part of that might well involve "daylighting" the lost rivers of London: bringing them back to the surface, and with them life and freshness, as cities around the world are doing—with benefits for the well-being of their citizens as well as the health of the environment.

Geoengineering is a literal attempt to make a better world than the one we have: it's the grandest of grand plans. It's no wonder people are wary of it, particularly those working close to the environment, to many of whom *more* technology is anathema. It seems more natural that we should get there by other means: by reducing our footprints, by rewilding the planet, by reconfiguring our industries. By tending to our gardens. And we *do* have to do that. But as Meehan Crist pointed out when asking if it is still right to have children, treating climate change as a matter of personal choice rather than systemic failure is misdirecting responsibility. We can *hope* everything works out, but that would be blind optimism. A planetary problem may ultimately require a planetary answer.

Think of it this way. We've *already* embarked on a vast experiment in geoengineering, but one that is virtually uncontrolled and caused by millions or billions of actors disposing of their waste in any way they wish, just as medieval Londoners were free to throw whatever they wished in "their" rivers. The effects are playing out both alarmingly quickly and terrifyingly slowly: even if we were to stop emitting carbon today, the effects of what we've already done will be felt for hundreds of years, just as the pollution of the Thames was felt for generations. The planet is our garden now, and its custodianship is ours, whether we like it or not. Or to put it in the words of the pioneering activist Stewart Brand: We are become as gods. We might as well get good at it.

9

Writing Tomorrow

What do we do about challenges we've never encountered before?

And speaking of gods, here comes artificial intelligence.

As the idea of voluntary human extinction was making its ascent from lunatic fringe to center-stage, another group of outsiders were also considering the potential demise of humanity. But they were more concerned with *in*voluntary extinction at the hand of external threats: meteorites, plagues, nuclear war and, above all, machines capable of outsmarting humans.

Nick Bostrom was a relatively obscure Oxford University philosopher until the publication of his 2014 book, *Superintelligence*. Superintelligent machines don't exist, nor is there a roadmap to build them, so it would be pushing it to call it a work of prediction: it's essentially a collection of thought experiments and logic puzzles also explored in the writings of Eliezer Yudkowsky, an autodidact who has fashioned himself into an artificial-intelligence maven via a series of gargantuan blog posts on the nature of rationality and cognition. Some of these scenarios describe how a computer might achieve and then rapidly exceed human levels of intelligence—with potentially disastrous consequences for its human creators. Most notoriously, Bostrom suggested that a superintelligent machine tasked with making paper clips might first find ways to commandeer all of the world's metal stocks, then all of its energy supplies; not satisfied with that, it might go on to figure out how to make paper clips out of pretty much everything else, until nothing would remain of the Earth and its inhabitants except one giant paper clip–making machine.

Superintelligence was not the sort of book that usually hits the bestseller lists. But it was well timed, being published just as AI, which had for some years lain moribund in what researchers called an "AI winter," was being reinvigorated. It turned out that what was needed to revive machine learning systems was vast quantities of data for them to train on—and the internet readily provided that. The most promising such systems were based on neural networks loosely modelled after the human brain—notably those built by DeepMind, a firm co-founded by Demis Hassabis, author of the scene construction hypothesis. Over the course of the 2010s, DeepMind's systems had aced a succession of tests that were supposedly impossible for machines to solve, starting with eerie, inhumanly successful ways of beating vintage Atari video games. A few years later, they had progressed to eerie, inhumanly successful ways of beating human grandmasters at the game of Go, another supposed human stronghold, and thereafter the hits kept coming.

At the same time, algorithmic working, particularly in the gig economy, was becoming cause for both hope and fear. While some were optimistic that this would lead to better utilisation of resources (spare rooms on Airbnb, private cars turned into Ubers), others worried about the potential for automation to put humans out of work—as many as 47 per cent of American jobs could be at risk, by one headline-grabbing estimate. There was plenty of scope to argue with those estimates, but plenty of scope for concern, too: even a far smaller percentage would still represent enormous disruption.

The combination of genuinely startling advances and generously sauced hype was enough to persuade many pundits that Bostrom's book might be more prophetic than philosophical. *Superintelligence* was endorsed by Elon Musk and Bill Gates; Bostrom's Future of Humanity Institute attracted millions of dollars of funding from the likes of Facebook co-founder Dustin Moskovitz. Yudkowsky's LessWrong website became the nexus of a large and active online community of self-declared "Rationalists" who debated the potential emergence of superintelligence, and ways to tame it, in enormous detail.

The risk of a superintelligence deciding to casually wipe out humanity was not the only so-called "existential risk" that shot to prominence during this period. Bostrom's Oxford-based Future of Humanity Institute had a counterpart, of sorts, in the Cambridge-based Centre for the Study of Existential Risk, founded in 2012 by Martin Rees, the Astronomer Royal and author of *Our Final Century*, and the philosopher Huw Price. CSER was concerned with several potentially extinction-level events: as well as runaway AI, its researchers were occupied with the existential risks we touched on in Chapter 7: bioweapons, pandemics, asteroid strikes and nuclear weapons, and sometimes esoterica like alien invasions and runaway nanotechnology.

Most of these "x-risks" are, to be clear, genuinely worthy of consideration, and in fact they have received such consideration for some time. The American writer Isaac Asimov listed many of these and more in *A Choice of Catastrophes* in 1979. Philosopher John Leslie published *The End of the World: The Science and Ethics of Human Extinction* in 1996; CSER co-founder Martin Rees's *Our Final Century* came out in 2003. What was new was the emergence of a community committed to tackling such risks—and of philanthropists, many of them minted in digital technology circles, who were willing to commit large sums to the cause: Moskovitz and his wife; Jaan Tallinn, co-founder of Skype; and, inevitably, Elon Musk.

It's not hard to argue that some of these areas should have attracted more attention in the past, but this avalanche of money has been of debatable value. Some previously niche areas, like biosecurity, risked being swamped by the cash, their priorities redirected to edge cases rather than core concerns. And scientists have to be trained, facilities built and research programmes designed: money can help with that, of course, but it takes time and expertise. Meanwhile, some areas, notably AI, saw a proliferation of organisations dedicated to talking and thinking, with outputs whose real-world relevance was not always evident. The advent of long-termism—the philosophy which implies that a vast number of future lives hang on our decisions—gave the gravy train more momentum still.

In other cases, there were more transparent and accountable risk-mitigation programmes being developed. In 2022, NASA spent $325 million to send a probe into deep space, on a mission which ended abruptly when it crashed at extremely high speed into an otherwise nondescript lump of space rock. Said rock was the humble asteroid Didymos, which had been minding its own business—but if it *had* been headed towards Earth, the devastation would have been on the same sort of scale as the asteroid that killed the dinosaurs. As it was, the impact successfully knocked it onto a new course. Should any newly discovered rock turn out to be on a real collision course with Earth (as one eventually will), we now have a credible way of dealing with it. So that's one existential risk down—without philanthropic money.

But the risk that attracted by far the most attention, and the most money, was the risk of human extinction caused by superintelligent machines, and in particular the problem of "aligning" such machines' activities with human priorities. To reiterate: no one knows how to build such a machine; the abiding belief was that at some point the machines would learn how to build their own increasingly intelligent successors, which they would do at incredible speed while still, for some reason, remaining extremely single-minded about jobs like making paper clips. To demonstrate this point, superintelligence believers would . . . well, they would do what we've seen over and over again: they would continue the "intelligence" line on a graph accelerating onwards and upwards. That the line might plateau, stutter or stall—as it has on previous occasions—was unthinkable.

Many x-risk believers *also* believe that appropriately aligned superintelligent machines could usher in a golden age for humanity. If a "fast take-off" event happened—a process also known as "foom"—super-smart machines would exceed our abilities before we had any chance of intervening in the process. But if well aligned, they would take a few milliseconds or so to make all the scientific breakthroughs needed to solve our earthly problems and ensure a life of plenitude for all. We might even decide to upload ourselves into them, for a post-human eternity in the Singularity, as Ray Kurzweil still claims will happen in the very near future. But how these rosier scenarios might play out attracts comparatively little attention either from the Rationalists or from the goggling public.

Nonetheless, their convictions grew stronger as the performance of the new AI systems continued to exceed expectations, often in ways unanticipated by their builders. Debate raged as to whether the machines were showing flickers of consciousness and independent thought, or were just "stochastic parrots" mindlessly emulating human behaviour. With every step, excitement mounted over their potential applications—but so did concern that they would be misused (to make deepfakes, say), or even escape human control altogether. Many of the scenarios—but especially the fearful ones—were discussed in meticulous detail, despite being almost entirely fictional: the philosophical equivalent of spooky stories told around a campfire.

Consider the example of Roko's basilisk, a hypothetical superintelligence which would be generally benevolent, but make a point of torturing those who had been aware of its potential existence but hadn't done everything they could to help it come into existence. Why would it do this? Because every day that it didn't exist would be a day in which many people would endure sufferings that its godlike capabilities could have prevented. So the bystanders' suffering would be justified by the encouragement it provided to everyone else. (In fact, it wouldn't punish the bystanders themselves, who would presumably be long dead, but ultrarealistic simulations of them.) And if you've just read this paragraph, you are now aware of Roko's basilisk, and thus your future simulacrum is doomed if you don't get cracking on artificial intelligence. My apologies if you had other plans.

To be charitable, this is a version of Pascal's wager, which suggests that we should act as though God exists because we lose little if he doesn't, but stand to gain eternal bliss if he does. (Error management!) If we are not feeling so generous, it's an overengineered version of those chain letters which instruct you to forward them to ten of your friends on pain of death. Defenders of the LessWrong community, where the Basilisk was first posited, argue that the uproar it caused—including Yudkowsky's response that posting such an "information hazard" was the action of a "fucking idiot"—was overblown and no one really believes it. But it illustrates the point, albeit in an extreme case: a "logical" conclusion has been drawn from an entirely fictional premise. Such arguments are intellectual shadow-boxing, not meaningful responses to the actual challenges of AI.

That matters, because while this intensely overheated debate was going on, the real and present dangers of AI were being swept under the carpet. Machine learning systems are biased if their training data is biased, as it usually is. Put "Editor-in-chief of a science magazine" into an image-generating AI, and it produces a wild-haired, ageing white male in horn-rimmed spectacles; as I write this, the editors-in-chief of four of the world's most prominent science magazines are all middle-aged women. That's irritating, but bias is much more of a problem if AI is being used to screen job applicants, say, or predict where a crime is likely to occur. AIs also hallucinate: since what they're doing is essentially an *extremely* sophisticated version of the predictive text on your phone, they don't always make the correct prediction. But they do so with great assurance: the potential for them to "poison the well" of information is worrying. The actual alignment problem, inasmuch as there is one, is between the values of civil society and of those of the corporations whose valuations depend on coming up with the next big thing—even if it doesn't actually work as advertised.

So AI doomerism would be somewhere between irrelevant and amusing were it not for the amount of influence these communities have gained. The runaway superintelligence narrative is perhaps the ultimate expression of the idea that technology is a force of nature, that it *wants* things, rather than being the product of human ingenuity and engineering. While this is a belief sincerely held by many working in AI, it is also a convenient narrative for big tech companies, because it allows them to claim that only they have the resources to corral such powerful technologies, and that government should let them get on with it—a message that resonates with the libertarian sentiments that echo around Silicon Valley and, thanks to vigorous lobbying, has found its way to the ears of policymakers and regulators.

In the absence of any substantive advance on alignment—hardly surprising, given that the subject of concern doesn't yet exist—the discussion has become pathological. Doom-mongers call for machine learning systems to be treated like weapons of mass destruction, subject to moratoria on their development, and with breaches taken with concomitant seriousness; in a *Time* op-ed, Yudkowsky called for rogue AI experimenters to be shut down by air strike if necessary. Singularitarians call for massive funding to bring about utopia: OpenAI boss Sam Altman reportedly sought $7 *trillion* of investment to buy colossal computing power for his machine learning system; he also mused that AI's vast energy needs might be met by nuclear fusion—a Holy Grail technology in itself, deemed just a means to the superintelligent end.

So: Will superintelligent machines lead to our extinction, or our salvation? Should we be ultimately optimistic, or ultimately pessimistic, about their arrival? Despite all the efforts of the x-risk community, we can't reason our way to an answer. We can all make educated guesses—there is much Rationalist debate as to plausible values of "probability(doom)" and "probability(foom)"—but they're ultimately just that: guesses.

But then again, some people make better guesses than others.

Super-Talented Weirdos

The world is not predictable, and neither are people. That's true for the behaviour of people in groups—which is why it's so hard to beat the stock market, or to predict the outcome of an election. But it's truer still when it comes to the actions of an individual: an autocrat deciding to invade a neighbouring nation, say.

There are *some* clues to such developments: the geopolitical context, troop build-ups or a newly belligerent tone. These tend not to be the kinds of signals that computers are (yet) adept at interpreting. That requires the human brain's ability to conjure up scenarios based on sparse data—filling in the gaps and drawing out possible outcomes. That's still not easy to do with any degree of reliability, which is why catastrophic events like the 9/11 attacks happen even when there turns out to have been plenty of intelligence ahead of time. But where machines fail, our minds may yet succeed—as long as they're the right kind of mind.

"If you want to figure out what characters around Putin might do, or how international criminal gangs might exploit holes in our border security, you don't want more Oxbridge English graduates who chat about Lacan at dinner parties with TV producers," wrote Dominic Cummings, Boris Johnson's consigliere, in a highly unorthodox job advertisement in January 2020. The people he wanted on his crack team of policy advisors instead were "super-talented weirdos," "wild cards" and "superforecasters."

That was probably the first the general public had heard of superforecasters—people who can predict political or economic events with remarkable accuracy—although they had actually been identified some fifteen years earlier by the political scientist Philip Tetlock. Tetlock's interest in prediction had been piqued by his participation in a 1984 effort to gauge the risk of nuclear war breaking out between the superpowers. He was struck by how confidently both hawks and doves stated their positions regardless of whatever evidence the other side produced to contradict then.

There had to be a better way to make such "expert political judgements," Tetlock decided. If he could get experts to make clear, precise and verifiable predictions, he might be able to determine the factors that contributed to their accuracy (or lack of it)—meaning not so much the information they used as the ways they made use of it.

To that end, he asked hundreds of "card-carrying experts whose livelihoods involved analyzing political and economic trends and events"—government officials, academics, journalists(!), and so on—to make tens of thousands of predictions. The topics included run-of-the mill subjects like domestic economic policy and performance, defense spending and geopolitics, and also more specific questions like the likelihood of a Gulf War (in 1990), the prospects of post-Communist states (1991-92) and the dot-com boom (1999).

Tetlock found that these superforecasters were much less fixed in their thinking: they were willing to "subvert" their own reasoning by considering alternative ways that things might play out—in other words, to take counterfactuals seriously rather than believe in some kind of hidden order or fate. Counterfactuals, Tetlock said, are "important diagnostic tools for assessing people's mental models of the past, and those mental models very much connect to how they think about possible futures." In other words, superforecasters are much more open to updating their beliefs when presented with information that undermined (or reinforced) their previous views.

To make the analysis more manageable, Tetlock framed his questions so that they could typically be answered by thinking about just three possible futures—"bad," "average" and "good" outcomes. To make sure that the answers could be objectively assessed, he asked for them to be phrased as numerical estimates: "30 per cent" rather than "quite unlikely." And he asked the experts to estimate how confident they were that their predictions were correct, borrowing a technique from meteorology to ensure that their final score depended not only on how accurate their predictions were, but on whether they had expressed an appropriate level of confidence in them.

What Tetlock found, after decades of asking and waiting, was that the vast majority of these "expert" predictions were worthless. In fact, the more acclaimed these experts were in their professional lives, the more worthless their predictions. Why?

Tetlock leaned on the philosopher Isaiah Berlin's division of writers, and thinkers more generally, into "hedgehogs" and "foxes." Hedgehogs know one big thing—which is to say they believe the world can be explained in terms of single, over-arching theories, like Leibniz's optimism, or Hegel's progress. Today, that makes them successful members of the commentariat, where big ideas, delivered with certainty, sell well. But in Tetlock's study, their predictions were no better than chance. That perhaps helps to explain the disenchantment with "experts" in recent years: despite exuding confidence, they get things completely wrong all the time. Foxes, on the other hand, believe a variety of approaches are needed to understand and engage with a complicated world, making them less appealing to soundbite media. But they did somewhat better in making accurate predictions.

Both groups were far outstripped, however, by a small group of people—around one or two in a hundred—who were strikingly more successful than the supposed experts, despite having no real expertise in the subjects under consideration. Tetlock found that these superforecasters were much less fixed in their thinking: they were willing to "subvert" their own reasoning by considering alternative ways that things might play out—in other words, to take counterfactuals seriously rather than believe in some kind of hidden order or fate. Counterfactuals, Tetlock said, are "important diagnostic tools for assessing people's mental models of the past, and those mental models very much connect to how they think about possible futures." In other words, superforecasters are much more open to updating their beliefs when presented with information that undermined (or reinforced) their previous views.

This last point is critical. It's in line with an approach to prediction known as Bayesian analysis, named after its inventor, the eighteenth-century theologian and mathematician Thomas Bayes. The basic process is simple: it's essentially the plot of any detective story. You start with a hunch about what's happened, or is going to happen—this is your "prior," in Bayesian-speak. Professor Plum is dead in the library, and Miss Scarlett seems nervous. Did he meet his end at her hands? Then you look into the matter more closely, gathering evidence, or "data," that helps you to refine your guess. There's a candlestick with blood on it; and Colonel Mustard's fingerprints are all over it. Now you're pretty sure whodunit: this is your "posterior." And it's time to call everyone into the drawing room to find out if you were right.

In principle, that's pretty simple. If it were that simple in practice, detective stories would be pretty dull. In fiction, and in real life, it can be much more complicated to come up with a good prior; there are lots of red herrings and secrets which make the job of gathering evidence more difficult; and the outcome may not be cut and dried. And if we're talking about a prediction, then the situation may well be fluid, and the data subject to change at any time. Hence the importance of being willing to change your views—to update your priors. That's another reason experts may not be good at prediction: if you spend a lifetime developing a view, it's inevitably pretty hard to change it.

So it might be an advantage to have no previous attachment to the topic under investigation, but be able to pick up and process information about it readily. Some people seem to be naturally good at this. Most of us are not. So Tetlock's Forecasting Research Institute (FRI) has since moved on to "tournaments" in which participants first make individual forecasts, then discuss them collaboratively in teams. Next, they discuss and scrutinise the forecasts made by other teams and update their own predictions accordingly. What makes this potentially valuable is the sharing of information, which you might expect to help all concerned to converge on a single answer.

In 2023, the FRI put together a group to discuss the risk of human extinction by artificial intelligence. It had previously found that AI experts tended to worry that the machines might get out of control and wipe us out. They put the odds of such a catastrophe befalling us before the year 2100 at 25 per cent. Generalist superforecasters, on the other hand, put the probability at just 0.1 per cent—250 times lower. Tetlock hoped that putting the two groups together might result in them trading expertise for perspective, sharing information and coming to some sort of consensus. That didn't happen. After hours of research and discussion, the gulf had barely narrowed. The experts had revised their estimate down to 20 per cent; the superforecasters had tweaked theirs up to 0.12 per cent.

Why didn't they change their minds? The two groups didn't actually differ very much on the short-term outlook for AI, or on the key pieces of information that would change their minds. But over the longer term they had very different ideas about what was important. The optimists felt that the prospect of human extinction was so outlandish as to require much greater evidence than was available. The pessimists, on the other hand, thought it was the emergence of an entirely new form of intelligence that was the extreme event, and that human extinction was one of the more plausible outcomes, just as the rise of *Homo sapiens* had meant extinction—whether deliberate or inadvertent—for other ancient species—including other ancient humans.

In other words, they were working from dramatically different priors. Even among people who know AI inside out, there are profound differences of opinion. Two of the three so-called "godfathers" of AI, Geoffrey Hinton and Yoshua Bengio, claim it poses an existential risk; the third, Yann LeCun, laughs that possibility off. So it's possible for people to come up with ideas about how this will play out that are wholly divorced from each other, and there's not enough hard evidence to bring their positions together.

And that in turn means that they cling to a preferred narrative: perhaps borrowed from history, perhaps inspired by fiction. If we want to get better at predicting and shaping the future, we need a longer list of narratives and a fuller list of priors. Coming up with those doesn't seem, on the face of it, to square with qualities common among superforecasters: numeracy, logical thinking and an almost mulish dedication to objectivity. What we need more of are imaginative flights of fancy—and the willingness to take them seriously.

The Law of Three Laws

Back in 1990, Mike Godwin, a lawyer with a keen interest in online matters, formulated a law of online discussion that later took his name: "As an online discussion grows longer, the probability of a comparison involving Nazis or Hitler approaches 1." When I was editing *New Scientist*, Godwin's Law inspired me to formulate a similar law involving artificial intelligence and the Terminator movies.

"As any online discussion of AI grows longer, the probability of a comparison involving the Terminator approaches 1."

In James Cameron's original 1984 film, the Terminator is a merciless android sent back from the future to ensure that Skynet, a superintelligent machine, ends up ruling the planet and exterminating humanity. It was by no means the first fictional depiction of such a subject, but one of the most memorable, continually renewed by an endless succession of sequels. Any story we posted about applications of AI, no matter how innocuous, would inevitably result in either an invocation of Skynet or a picture of the Terminator. It didn't usually take very long, either. This was particularly true of stories that hinted at AI with unexpected capabilities, regardless of whether these were remotely threatening.

There was an alternative, one frequently cited by the makers of AI systems: that if a system *did* pose risks—usually closer to the "breaking things by accident" end of the spectrum than to "genocidal mania"—it could be brought under control using a set of rules like the Three Laws of Robotics, formulated by Isaac Asimov in a legendary series of short stories in the 1950s. The laws stipulate that a robot must not harm humans, must obey human orders and must protect its own existence. Asimov used them to set up logic puzzles thinly disguised as stories, normally revolving around a robot whose baffling behaviour could be traced back to some situation that had created some irresolvable contradiction between the Three Laws. They are not, in fact, so very different to the stories the AI x-risk community tells, although the characterisation is marginally more convincing.

These stories should themselves really have been enough to demonstrate that the Three Laws are completely useless, as Asimov himself acknowledged late in his career. It's possible to come up with endless scenarios in which the letter of the laws is observed, but their spirit is trampled into the dust. Understanding the spirit of the laws, on the other hand, would require machines with a detailed conceptual understanding of our (actual) world that remains a long way off. Robotic vacuum cleaners don't obey the Three Laws of Robotics: they just don't go very fast and they're designed not to chew up fingers. Self-driving cars don't know they're not meant to harm people, or even what people are; they're just trained to avoid running into things. That's why they can be foiled by poorly placed traffic cones—or by kids jumping in front of them.

Most people who actually work on AI or robotics are perfectly aware of this, but nonetheless cite similar sets of rules as design principles, rather than control systems. In 2011, for example, two UK research councils proposed five ethical principles and seven "high-level messages" for those working on robotics, the first of which read "Robots should not be designed as weapons, except for national security reasons." (Skynet!) In fact, this was highlighting that robots, like any guns or knives, have both productive and destructive applications: the onus is on the designer and operator to ensure they're productive. And it's actually a simplified version intended for general audiences; the full version is more nuanced.

Both the Terminator and the Three Laws are recent manifestations of a very old story: its first iteration is perhaps the legend of the golem, a humanoid sculpture made out of clay and animated by a spell. In the canonical version of the legend, there is an error in the spell—its "programming"—and the golem runs amok until it is corrected. Much the same story is retold in Goethe's 1797 poem *The Sorcerer's Apprentice*, adapted as the sequence in *Fantasia* in which Mickey Mouse loses control of his magically multiplying mops and buckets; in 1818's *Frankenstein*, in which, of course, a reanimated patchwork corpse defies its creator's instructions, to lethal effect; and in *R.U.R.*, the 1920 play which gave us the word "robot"—and a robot rebellion.

The recurrence of such stories—which clearly don't refer to modern conceptions of artificial intelligence—suggest they have more to do with our hopes and fears of technology than the technology itself. Kanta Dihal investigated the way public perceptions of AI are being shaped while at the Leverhulme Centre for the Future of Intelligence at the University of Cambridge; her survey of the British public revealed that most people took their cues from only three films: *The Terminator*, Steven Spielberg's *AI*, about a robot child cast out of his adoptive family, and *I, Robot*, another ominous tale despite being loosely adapted from an Asimov story. "It's a very narrow set of narratives that keep being reused and that narrowly steer people's expectations," she observed.

Such Hollywood narratives are now being pushed out globally, where they may conflict with existing narratives about intelligent machines. So Dihal and her project team ran workshops to investigate how intelligent machines were perceived all over the world. That revealed notable differences: Japan, for example, has long depicted robots as friendly and helpful. "While many of us have fears of AI taking jobs, those concerns are much more limited in many other parts of the world, for instance in Japan, and to a lesser extent in South Korea. There's more emphasis on care robots being developed in order to support an increasingly aging population, which is of course a very different situation from say, India, where technological unemployment is a much more pressing issue."

It may seem odd to talk about a lack of imagination when we started out talking about godlike machines turning everyone on the planet into paper clips. But the anxiety about superintelligence stems from just that. The conversation about AI, at least in the Anglosphere, has been ceded to a small group of people who pride themselves on their rationality while conforming to fixed views about how the technology will develop, and who pay minimal attention to its societal context. Rather than attempting to understand the myriad outcomes that are possible—and which are within our control—it is dominated by the sense that we are powerless to stop the machines forging their *own* destiny, a sense which owes everything to fiction and very little to the real world of software engineering.

It's customary for those working on AI to be treated as seers, their proclamations to be received as bold visions of a future that the rest of us can barely envisage. But in fact what's remarkable is the *poverty* of imagination on display. Faced with the prospect of a truly revolutionary technology, they—and we—simply fall back on the same stories we've been telling for hundreds, if not thousands, of years.

Bowing Down Before Reality

Catastrophising technological progress is not even remotely new. In 1859, Charles Baudelaire, the poet, critic and debauchee, warned that the advent of photography threatened to "ruin whatever might remain of the divine in the French mind." Painting allowed expression, he argued, whereas photography offered only exactitude: "Each day art further diminishes its self-respect by bowing down before external reality; each day the painter becomes more and more given to painting not what he dreams but what he sees." But the new technology nonetheless won out, thanks to what Baudelaire called the desire of "squalid society" for cheap and pornographic spectacles. Painters would be put out of a job—apart from those venal enough to become photographers.

It's difficult to read these words today and not be reminded of concern about online media, where moments of great beauty and wonder are presented alongside what might fairly be called cheap and pornographic spectacles. More recently, generative AI has proven able to produce passable images at the blink of an eye, raising concerns about the cheapening of art, the potential for corrupting imagery—with the rise of deepfaked pictures and videos—and the forced redundancy of artists.

Go looking, and you can find jeremiads about pretty much any technology you care to mention: memory-sapping writing, head-turning novels, mind-rotting television, teen-addicting games and eventually the internet. There are sites and feeds, such as the Pessimists Archive, that specialise in surfacing the histrionic and frequently hilarious cautions that have been offered in the past—why bicycles would lead women to lose their morals, how elevators would cause debilitating "brain fever"—but despite all the hand-wringing, civilisation has endured. Thrived, even. Or has it?

There are two ways of interpreting such warnings. One is to imply that since these warnings proved unnecessarily anxious in the past—and often, with the benefit of hindsight, seem positively ridiculous—they're unnecessary now, and will come to look equally ridiculous to those who grow up with them. We need not worry about mental degradation, social corruption, mass unemployment or other ills associated with new technology because we've been here before, and in the event things turned out just fine.

This view, which we might call the complacent perspective, holds that people are fundamentally good at figuring out how technology can be made to serve their purposes, and not vice versa. Writing wasn't the end of memory; it just turned it into a new and more useful form—one that allowed knowledge to be shared no matter what happened to our own fallible minds. Photography didn't stunt our imaginations; its exactitude allowed us to record sights we could never capture, and in due course we found ways to use it expressively, too. Sure, some people didn't like it, and some people lost their jobs. But in the end the new technology became the old technology, everyone got used to it and life went on.

The despairing perspective is that perhaps Baudelaire had a point. Civilisation might not have collapsed, but that doesn't mean that we haven't lost out along the way—in fact, we might not realise how much we've lost. Like Baudelaire's photo enthusiasts, we are so gratified by the vulgar charms of a new technology—whether sensational media or labour-saving machinery—that we don't stop to consider whether its benefits are really worth having. Over time, we gradually lose our faculties and our well-being, becoming mere lumpen imitations of our former, nobler selves.

This is, after all, what has happened with, say, food. Before the introduction of sugar, dental disease was relatively uncommon. But over time, sugar became added in increasingly large proportions to our food—and so did any number of other additives that made it cheaper and tastier, but more calorific and less nutritious—until we are now afflicted with a host of ills, from diabetes to creaky hips, that arise solely from the disguised crappiness of what we eat. We have managed to create palatable diets that result in both obesity and malnutrition. You *could* argue that the same has happened to our media diet, with increasing consumption of junk information damaging our mental and social health.

Both the complacent and despairing perspectives have merit; both can be pathological in excess. The truth lies somewhere in the middle. In 1986, the historian Melvin Kranzberg formulated six "laws" of technology, which I think of as a kind of sociological parallel to Asimov's technological rules. The first states: "Technology is neither good nor bad, nor is it neutral." We should be neither complacent nor despairing of technological interventions in our lives, but we should not just let them run their course either—particularly as they get so powerful as to be genuinely world-changing.

Previous world-changing technologies have been subject to controls of various degrees: the pesticide DDT and ozone-destroying CFCs were phased out when their environmental effects became known. Leaded petrol and asbestos were similarly abandoned when their toxicity became apparent. We have procedures for testing the safety of new medicines, and safety standards for everything from children's clothing to aircraft doors. Research into biotechnology has been effectively governed by principles laid down at a conference in Asilomar, California, in 1975—a venue and model emulated by some AI researchers in 2017. There are ways to ensure we strike a balance with any new technology. But the digital world seems to have caught us rather off-guard.

We've seen over the past two decades what happens when technologies, and technologists, are allowed free rein. The results are neither good nor bad, and they are decidedly not neutral. Allow an appified cab company to scoff at taxi regulations, and you end up with cheap and convenient car services; you also end up with exploited drivers, damaged roads and air pollution. And when the investment tap runs dry, you lose the cheap and convenient car service, but your regular cabbies have all quit. And Elon Musk's robotaxis are still nowhere to be seen.

Today's billionaire technologists wield wealth and power which superficially rival that of Gilded Age oil and banking tycoons—but operate in very different ways. Digital products and services can be reconfigured with far greater ease than physical assets and infrastructure; network effects are more easily established; and it's easier to keep one step ahead of whatever checks and balances might be put in place. But we don't have to accept that technology "wants" anything, or for that matter that technologists should be allowed to do whatever they want either. Technology does not make itself, at least not yet: we do. We can stop making it, we can make it faster, or we can make it more carefully. The choice is ours, if the wit and will exists. And to make those choices *before* "something has hit us hard," as H. G. Wells put it, we need to tell the right stories about the technologies, and societies, that we actually want.

Steam Engine Time

Put a sufficiently compelling narrative into the world, and someone will try to make it come true. It's not impossible to anticipate a particular technological development, even if you have no idea precisely how it can be achieved. You can do it by watching technological trends and cross-referencing them with what people want or need.

In 1982, the writer William Gibson was struck by the poster for an Apple computer—a poster which made it look like a piece of consumer hardware, not an industrial machine. "Everyone is going to have one of these, I thought, and everyone is going to want to live inside them. And somehow I knew that the notional space behind all of the computer screens would be one single universe," he explained.

That led him to coin the term "cyberspace": "A graphic representation of data abstracted from the banks of every computer in the human system. Unthinkable complexity. Lines of light ranged in the nonspace of the mind, clusters and constellations of data. Like city lights, receding." At this time, remember, the entire "internet" was comprised of a network called ARPANET that connected a few hundred academic and military computers. There was no graphical interface, no navigation or links: everything was done through what amounted to snippets of code. Gibson wasn't online; he didn't even have a computer: He wrote his stories on a manual typewriter.

"When I came up with my cyberspace idea, I thought, I bet it's steam-engine time for this one, because I can't be the only person noticing these various things," he wrote. By "steam-engine time," Gibson was referring to the tendency for the same theory or technology to be invented by several people independently at the same time. "And I wasn't. I was just the first person who put it together in that particular way, and I had a logo for it, I had my neologism." Gibson's first novel, *Neuromancer*, was revolutionary; he was soon joined by other "cyberpunk" writers who converged on a vision of the online world as a mildly dystopian but radically permissive space—constructed by giant corporations but inhabited by hackers and outlaws—that proved enormously influential in shaping the culture of the internet.

Writers have long tried to exploit this narrative power for good; remember Edward Bellamy's *Looking Backward*. During science fiction's pulp era, many authors believed they were writing the future into being: one influential group was the Futurians, founded by communist-leaning editor Donald Wollheim, who believed "science fiction followers should actively work for the realization of the scientific world-state as the only genuine justification for their activities and existence." Innumerable technologists have claimed inspiration from *Star Trek*, which has been credited with many inventions that first appeared in the pulps—including personal communicators (which manifested as flip-phones) and replicators (3D printers).

Science fiction "needs to be understood as a kind of modelling exercise, trying on various scenarios to see how they feel, and how deliberately pursuing one of them would suggest certain actions in the present," wrote genre legend Kim Stanley Robinson in a 2016 think-piece for *Scientific American*. Robinson wrote a classic trilogy about turning Mars into a habitable planet; his much-praised 2020 book, *The Ministry for the Future*, is almost a manual for renewing the habitability of the Earth, complicated by rogue geoengineers and extinction terrorists, facilitated by degrowth and international cooperation. Now there's an attempt to establish precisely such a "ministry for the future" at the Oxford Martin School of Business. "It is at one and the same time an attempt to portray a possible future and an attempt to describe how our present feels," Robinson continued. "The two aspects are like the two photographs in a stereopticon, and when the two images merge in the mind, a third dimension pops into being; in this case, the dimension is time. It's history, made more visible than usual by way of science fiction's exercise of imaginative vision."

It's this split—reflecting on the past, imagining the future—that makes science fiction such a potent vehicle for our hopes, dreams and fears. There are numerous strands of fiction today—not just written, but across all art forms—that explicitly set out to inspire particular kinds of innovation. "Solarpunk," for example, attempts to depict the clean beauty of a post-carbon world in which gardening and grand designs are combined.* It's primarily visual, as is "Anglo-futurism," a flag-waving twenty-first-century remix of Bazalgettian mega-engineering. "Speculative designers" go one dimension further, making objects and experiences that allow people to explore their reactions to technologies and scenarios that don't yet exist: for example, a sniff of air samples from the future, laden with pollution or scrubbed clean and fresh. Fake it till you make it.

What inspiration people take from such exercises can't be assumed, of course. Cyberpunk inspired a legion of digital technologists, but as Gibson told me in 2020, "I could never understand where their optimism came from, when people started to speak to me of disruption with what looked like delight. I don't think it was the guile of greed, there was just faith that it would be okay . . . and they hadn't seemed to notice that the world of *Neuromancer* was fairly problematic."

That process continues. One of Hollywood's most refreshing, post-*Terminator* depictions of AI came from the 2013 film *Her*, in which Scarlett Johansson plays a disembodied AI who provides friendship to a lonely man—which proves to be a bittersweet experience. A decade later, "virtual girlfriends" had become a real-life talking point, and *Her* also became relevant for another reason: OpenAI was forced to remove one of the voices for ChatGPT because it was uncannily similar to Johansson's, who had repeatedly declined to provide hers. Narratives can be used, abused and co-opted: to paraphrase Melvin Kranzberg, they are neither good, nor bad, nor neutral.

* Rather bizarrely, one of solarpunk's most lauded works is *Dear Alice*, a short animation advertising yoghurt. Cows chew the cud under solar panels; tentacular robots harvest fruit. Take inspiration where you find it.

One reaction to that has been an emphasis on collaborative ways of creating visions of the future. Stephen Oram is one of several writers practising "applied science fiction," working with scientists and technologists to devise credible visions of the future that can be shared with the public. Veteran futurist Julian Bleecker hosts online conversations that aim to foster ideas about "everyday AI"—when the technology is so established as to have become boring—developing a bottom-up picture that contrasts with the hyperbolic projections emanating from the tech industry. Community workshops and citizens' assemblies aim to provide people with the information and ideas needed to envisage their shared futures. One common format is the "postcard from the future," a riff on the "best possible self" exercise. You might like to try writing one to yourself.

Building Foundations

Technological innovation is shaped and spurred on by the stories we tell ourselves. But social innovation can be, too. In the Foundation books, Isaac Asimov's other classic series, the lone genius Hari Seldon develops a way of algorithmically predicting the future—not its precise details, but its broad arc. Using this "psychohistory," he foresees that the seemingly mighty Galactic Empire is in fact on the verge of collapse, but also devises a plan to reduce the subsequent period of barbarism to a mere millennium, rather than a dark age lasting tens of thousands of years, by establishing a "Foundation" of scholars, based on an out-of-the-way world far from the chaos, who will steer the course of events over the centuries to come.

Asimov began writing the series in the early 1940s, inspired by Edward Gibbon's epic *The History of the Decline and Fall of the Roman Empire* and the theory put forward by historian Arnold Toynbee that all great civilisations follow the same cyclical pattern. (It was Toynbee, you'll recall, who mocked the idea that history is just "one damned thing after another.") In particular, Toynbee suggested that it's up to a "creative minority" to solve the challenges their societies face, to the benefit of the unwashed masses. Asimov's series follows the Roman template, but with spaceships and blasters instead of chariots and swords. As his Galactic Empire disintegrates into myriad fiefdoms run by warlords, robber barons and gangsters, conflicts arise but are duly resolved through the Foundation's intervention.

Foundation is optimistic through and through: it's literally the story of how brainiacs learn to predict the future so clearly that they can act to ensure the best of all possible galaxies, without even breaking cover—like some sort of inverted conspiracy theory in which the Illuminati only have the best interests of the public at heart. Its reassurance that the seeming chaos of human affairs can be navigated is, I think, why it has retained its popularity for decades despite its pedestrian writing, flat characters, dated social mores and inherently anti-climactic plotting. It's no coincidence that most readers find Foundation in their teenage years—and no accident that some follow careers that parallel it.

"For a high school student who loved history, Asimov's most exhilarating invention was the 'psychohistorian' Hari Seldon," Newt Gingrich wrote in his 1996 political prospectus, *To Renew America*, published in the same year as his World War II counterfactual *1945*. In the book's index, "Asimov, Isaac" appears immediately above "Assault weapons"; "*Foundation* trilogy (Asimov)" appears above "Founding Fathers."

Gingrich, the architect of today's Republicanism, credited Foundation with awakening him to the potential for civilisational ruin. It also gave him the blueprint for averting it. In 2011, Gingrich announced he was running for president; Ray Smock, a former historian of the US House of Representatives, anatomised the parallels between Gingrich's career and Hari Seldon's. Gingrich and his fictional inspiration were both professors who had moved to the bureaucratic nexuses of America and the, er, planet Trantor respectively to warn of imperial decline, Smock observed. Where Seldon had his Foundation, Gingrich co-opted GOPAC, a Republican political training programme, to create a cadre of hardened operatives who could carry out his plan to reshape US politics to be far more ruthlessly partisan. (It seems unlikely that the avowedly liberal Asimov, who died in 1992, would have approved.)

Smock concluded his hit piece on Gingrich by quoting Cicero: "To be ignorant of history was to remain always a child. To which we might add a Gingrich corollary: to confuse science fiction with reality is to remain always a child." Perhaps so. But Gingrich's childhood inspiration and unrealistic expectations brought him from the depths of academia to the heart of power—where he successfully changed the future for decades to come.

Utopias have a tendency to turn dystopian. *Star Trek* is noteworthy as one of the few mass-media depictions of a utopia, a post-scarcity society in which conflict and labour have been replaced by boldly going to brave new worlds. At least, it was originally: but utopias are boring, and impossible, hence the need to liven up episodes by having Kirk perpetually either falling in love or getting into a fist-fight. Telling interesting stories while sticking to *Trek*'s utopian roots has proven ever harder as time's gone on: the tone of later series has become ever darker.

In fact, dystopia seems to have taken the upper hand in our imaginations in recent years: grim, post-apocalyptic worlds abound, along with cautionary tales of how it could all go terribly wrong. Perhaps that just reflects the mood of our times: consider the resurgence of Margaret Atwood's *The Handmaid's Tale* at a time when reproductive rights are being challenged and having children has become a source of angst. Atwood's cautionary tale was nominally set in the very near future when it was published in 1985. But nothing about the world depicted relies on new technology. Quite the opposite: Atwood has repeatedly pointed out that everything described in the book had happened at some time in recent history, somewhere in the world, as documented by newspaper clippings she collected while writing it. Her innovation was to construct an alternative United States where these things were practised by a version of the religious right. The real and fictional Americas have converged; and so has their symbolism. *The Handmaid's Tale* gave reproductive rights campaigners a common reference point—visually, too, with their white bonnets and red cloaks.

And what, then, of AI? If we *did* make a genuinely superintelligent AI, we have no idea what it would be like: It might be able to solve problems or create material faster than any human, but would its intelligence be anything like a human's, without a body, a social life or biological wants and needs? There are, after all, many superhuman machines already—unless you're a savant, the calculator function on your phone can probably do arithmetic far more quickly and accurately than you can, but we rarely worry that they will exploit their long-division skills to take over the world. AI neither wants anything in the sense that we understand "wanting," nor does it have any conception of the physical world.

And because we have no idea what a superintelligent machine would really be like, or what it would want, we don't have any real idea how likely it would be to run amok. It's a possibility, sure, but so is pretty much any other outcome. Why would such a godlike machine be unable to exercise common sense, or even be bothered about its original instructions? We might as well hypothesise that it would commit itself to the benefit of all life on Earth—or turn itself into an interstellar starship and abandon us to our own devices. Or any number of other things.

Right now, we are better off creating stories about the kinds of AI we would *want* to live with, and the kinds of societies that might result. We probably won't get them right away, but at least we can start down that road. We already have some choices. In Iain M. Banks's Culture novels there are a range of AIs, from simple drones to hyper-intelligent minds, which steward and safeguard humanity. We have companionable AIs, including the superintelligence in *Her*. There's the mischievous co-star of *Robot & Frank*, which subverts expectations of robot-carers for the elderly. The TARS helper-robots in *Interstellar* have programmable "humour" and "honesty" settings; Marvin the Paranoid Android from Douglas Adams's *Hitchhiker's Guide to the Galaxy* is mired in depression. All of these, and more, represent conceptions of how AI might play out.

And AI itself could be a powerful tool for generating new narratives. The visions and stories that AI can conjure up are uncanny remixes of human imagination: right now, however, it's mostly being driven to produce low-grade mimicry of human creativity. It would be a missed opportunity—a misalignment—if, like other fruits of the digital age, it becomes exploitative and monocultural rather than assistive and diverse. We may write our next chapter with the help of optimistic machines.

As we learn more about the real technology, those depictions will become more credible: fiction can be Bayesian, updating as we learn, too. That will help us to shake off our fixation with AI, or any other technology, either destroying or liberating humanity: with having to choose between catastrophe and cornucopia. The best of all possible worlds, the brightest of all possible futures, is ours for the taking. We just need to imagine them, believe in them—and start making them.

EPILOGUE

The Best of All Possible Universes

One of the most curious things about our universe is that it's astonishingly well-tailored to the existence of human life.

The consensus among cosmologists is that space and time came into existence about 14 billion years ago, in the event we call the Big Bang. As the universe expanded and cooled, it became filled with a range of fundamental particles; the laws of physics saw to it that some of those particles vanished, while others assembled into ever more complex structures: galaxies, stars, planets, lifeforms, us. But why *those* ingredients, and *those* laws?

The possibility that other ingredients and other laws might give rise to other kinds of universe has been much debated in recent years. Only in some universes where the combination of ingredients and laws are just right—for reasons that remain enigmatic—could intelligent life arise. It's even possible that we just happen to live in a bubble of spacetime where the laws sympathetic to human life prevail; out there somewhere, there might be other bubbles where the laws are entirely alien and inimical to our existence.

So there really might be many possible worlds. And by mere virtue of the fact that we exist, we're in one that suits human life. It seems too great a coincidence to accept blindly. How can this be?

One answer to this conundrum is that God, or some other force—call it *Kosmischegeist*, if you like—simply arranged things that way.

Another answer is that, actually, you can set the values differently and still come up with viable universes, if you just arrange them carefully. In those counterfactual universes, we would perhaps not be human as we are, but something else. Something else, but perhaps something that would still quest for meaning in an apparently indifferent universe.

And a third answer to the conundrum of our existence is that there are, in fact, many different universes, most of which *are* desolate places. We naturally exist in the one that permits us to exist because we couldn't exist in any of the others. Ours is the best of all possible universes.

Perhaps Leibniz had a point, after all.

And then there's the other kind of multiverse—the quantum many-worlds version, in which the universe branches every time you make a decision, or take an action . . . or something. No one seems quite sure what's necessary to make that happen, which is one reason I'm suspicious of the many-worlds interpretation. I think it maps too conveniently onto the way we already think about the world, the narrative structure we impose on reality.

Some people are convinced the cosmos works this way, or claim to be. Back in 2014, I edited a feature on the many-worlds theory for *New Scientist*. It wasn't about the scientific arguments, such as they were: it was about what it meant to live in the belief that countless other versions of you are out there somewhere, going about their own lives. We thought the best people to ask would be physicists, so we asked them.

One replied that when he'd been expecting his first child, he found himself hoping that it would all go well. But then he thought: "It was going to go well, and it was going to end in tragedy, in different parallel universes. So what did it mean for me to hope that it was going to go well?" You can't just wish your way onto a "good" branch of the multiverse. Like Einstein thinking that his best friend Besso had never really died. Or lived.

Some people *do* believe that you can jump from one world to another. Some believe it explicitly, trading off the garbled versions of quantum physics that are the nonsensical stock-in-trade of the alternative set, some more implicitly: ask and believe, and the universe will provide. There's a seduction to this idea, that you can find your way to the world that suits you best. You might believe that, if you can only stay positive enough, if you can only believe hard enough, you can jump to the one world in ten where you're cured of cancer.

Of course, the world doesn't work that way.

Another of the physicists we asked was a devout Christian as well as a many-worlds believer. He told us it solved the problem of evil. God values elegance above suffering, he told us, and so "God won't collapse the wave function to cure people of cancer, or prevent earthquakes or whatever, because that would make the universe much more inelegant." In his account, God determines everything, and humans have no free will—but we don't need it. Because of the multiverse, we do everything, everywhere all at once.

This, then, is where we've got to in understanding what we can expect of the world—with our grandest theories of how the cosmos is ordered. Hope is meaningless; God cares more for elegance than suffering.

I guess Voltaire had a point, too.

Hoping Against Hope

The distinction that's sticking in my mind right now is the one that can be drawn between optimism and hope. If someone has a one in ten chance of a positive outcome—of being successfully treated for cancer, say—someone taking an unrealistically optimistic view will think their own *personal* chances of being cured are actually higher than that, even if they know the odds overall. A hopeful person, on the other hand, understands the odds well, but wishes without expectation to be the one person out of ten who benefits.

Kathryn's cancer treatment team were, I think, running high on dispositional optimism. Embracing it and exuding it must surely be a requirement of the job. I think I was being hopeful. I knew the statistics; I merely hoped she would land on the right side of them. And Kathryn, I think, was being unrealistically optimistic. She trusted me on the statistics; she believed that she would be cured; and she clung to that belief through all the rounds of surgery and chemo and radio.

I don't think any of us knew we were working from different species of optimism. Perhaps it would have been too hard to differentiate between them anyway. But I think it might have helped if I had known. I might have been able to counsel her better. I might have been able to comfort her better. I might have been able to prepare better. I think it would have helped.

The moments after Kathryn died were, in my memory, still and quiet ones. I had wondered for months what it would feel like, but the answer was that it felt like nothing. I hadn't slept for days and I had exhausted every possible shade of emotion. So the answer was that I felt nothing: numbness and emptiness.

I didn't want to go home, so my friends took me to someone's house—I don't remember whose—where I slept and woke, the next day, feeling . . . positive? I felt as though whatever happened now, it would not involve the endless wearing down of hope. I knew that at some point it would hit me. I was waiting for the moment. Surely there would come devastation. But it didn't come.

Eventually it was time to go home. It was a sunny day, and the house was full of light. We all made a cup of tea. The conversation was flowing—I have no idea what we were talking about, but it wasn't what had just happened. Somewhere, I knew that they would eventually have to leave and I would have to learn to be alone in my home.

I slipped away and went up to our bedroom. The curtains were still drawn, so I opened them. A fleeting memory of the panicky, breathless scramble the last time we left the house. I did some busywork: put something away, took stock of the room. It smelt of her. I sat on the bed, and then, entirely without warning, I howled. Keened. It was unlike any experience I'd ever had, nor was it like any I've ever had since. There was no thought, only loss.

I still think about that moment often, still trying to puzzle out where my mind went. But I think the answer is that it hadn't gone anywhere. It just wasn't there. There was no way to reason through the fact of her absence: it was a reality that needed accepting, and not by the ratiocinating parts of my mind, which had seen it in the mortality tables in the peer-reviewed literature long before. It was the unreasoning, unreasonable part of me that had clung on to hope throughout that longest and shortest of years and which now, finally, had to let go.

In that moment, I realised—not articulately, not at the time—that the past had closed, was now cut off from the present. That period had ended. There would be no continuity with what came next. There would be no returning home. My house was still there; my home was not. But after a year during which branches of the future had been systematically pruned away by the progress of the disease, there were, abruptly, many new paths again. And now it was up to me to choose one.

From Extinction to Simulation

"The term 'technological fix' has become as pejorative as 'Luddite' used to be. The aspiration for technological solutions is now widely regarded as naive, a fantasy that ignores the inevitability of missteps and side effects. And that naivety is labelled optimism," the physicist David Deutsch said at the Royal Society for Arts, Manufactures and Commerce in 2015; he was representing the side of optimism in a debate with professional pessimist Martin Rees, the founder of the Centre for the Study of Existential Risk.

"Conventional pessimism is correct to say that civilisation has no guaranteed future, nor does our species," Deutsch continued; after all, almost every civilisation and every species that ever existed has become extinct, including our own relatives and ancestors. And that was because they hadn't known enough to stay alive: they didn't have germ theory, for example, so they suffered devastating pandemics. If they had, he suggested, we would have advanced so quickly that "technology would be regulating trivialities like the planetary climate as automatically as it's now regulating the temperature in this room."

"Every transformation of physical systems that is not forbidden by laws of physics is achievable, given the right knowledge, and hence the rational attitude to the future is what I call optimism," he concluded. "The principle of optimism, namely, that all evils are caused by lack of knowledge." The answer, he suggested, was to press on with discovery; while there would be side effects, it's impossible to know what they will be, just as no one knew in 1900 that uranium would be the centrepiece of existential risk, or that carbon dioxide would be today. We'll deal with them as they come. A very Odyssean approach.

Deutsch is, as it happens, a staunch advocate of the reality of the multiverse; he regards the evidence that it exists, from subatomic physics, to be as compelling as fossils are to the existence of dinosaurs. I'd love to say he found common cause with Leibniz's optimism and his array of possible worlds; but in fact he gives his German predecessor short shrift in his book *The Beginning of Infinity* for relying on God to fill the hole in his argument. I can't say I really disagree, but I think possible worlds has use as a metaphor for optimism—maybe there's a book in it? And for that matter, I think the multiverse is no more than an elaborate metaphor either, for cosmological truths that may forever be incomprehensible to humans.

There's a new metaphor on the way now, though, one that is taking hold just as quantum uncertainty did in the mid-twentieth century, and the multiverse has over the past three decades: the idea that we all live inside a giant computer simulation, built by beings who have already achieved supreme mastery of technology. The reasons that this idea has become so popular in some quarters are almost embarrassingly obvious, in a world obsessed with its own, ongoing digital transformation. But it worries me. In the multiverse metaphor, anything is possible; decisions and actions determine which world we live in. In the simulation hypothesis, we are just programmed to do what we're told: a return to determinism, to fatalism. To tragedy.

New Scientist, Manifesting and Me

When I was a kid, I sometimes went up to my dad's offices during the holidays. There wasn't an awful lot to occupy a pre-teen there, but I was well used to passing the time with my nose in a book and my head in the clouds.

One day I stumbled across a stack of *New Scientist* magazines. This was a novel and exciting concept: I liked science, and I liked writing, but was dimly aware that these were not widely regarded as compossible life skills. I'd be lying if I said *NS* was kid-friendly, but its text was readable and its cover art amazing—bright, dynamic and imaginative. Its pages were stuffed with discoveries and inventions; at the back there was a comic strip and a column by a mad scientist. The particular cover that caught my eye—at least, as my memory reconstructs it—was for a special issue investigating the paranormal, featuring a skull, a UFO and a bent teaspoon.

This, I decided, was what people who liked science *and* writing did for a living. And that would be what I'd do when I grew up. I asked my dad who took care of a magazine like *New Scientist*. The editor, he said. So I decided that was the job I'd do.

I kept reading *New Scientist*—first hand-me-downs from my father, then copies at school, and eventually a subscription I won in a competition judged by another future editor. In due course I went off to university to study physics; then I switched back to writing. By now it had dawned on me that being the editor of *New Scientist* was not a job you just waltzed into. In fact, getting *any* job there seemed like a stretch.

But I lived in hope. So I wrote to the features editor and asked for a job, on the basis that I liked science and I liked writing. He wrote a kindly letter back, suggesting that I should get some experience first. I applied for the first job in journalism that I could get, which turned out to be for a boring financial magazine. But I worked with other youngsters, it paid the bills, and I got to see the world while doing nothing more arduous than cranking out words. I fell in love, got married—and then everything changed.

While I was picking myself up, I saw an ad for a temporary job at *New Scientist*. Having worked in finance for more than a decade, I couldn't see how I was suited to the job. But I was trying to find my best possible self, and I had nothing left to lose, so I applied. And I did get the job, from the man I had written to, working with the man who'd awarded me the subscription. Working at *New Scientist* proved to be both just like I'd expected, and absolutely nothing like it; but in any case, one thing led to another and I ended up, some thirty years after my first acquaintance with the magazine, serving as the editor.

Does this testify to the value of optimism? Perhaps it does: a self-fulfilling prophecy enacted over decades. Or to put it another way, there were any number of points at which I could have succumbed to a pessimism trap. You miss every shot you don't take.

Or perhaps I could have got the job much sooner if I'd been *more* optimistic. In another world, perhaps I did; but I don't have access to those worlds, just this one.

Perhaps in another world I'd take a different, more mystical message from that special paranormal issue of *New Scientist* and I'd call it manifesting: I wanted something, and in due course the universe provided. Of course, memory is not a reliable narrator: it would be no surprise to discover the account above is inaccurate. There's explanatory style: the account undoubtedly ticks the optimistic boxes. There's confirmation bias: this is the coincidence that happens to agree with my priors. Or my subconscious at work, constantly steering me in the right direction.

Does it matter what we call it, as long as it works? I think it does and it doesn't. It matters when people more important than the editor of *New Scientist* blunder on with their own quixotic versions of optimism. It matters when it comes to how we make collective decisions: a society can't run on whim and wonder alone. But when it comes to how you live your life, your one and only life, the one here and now? I hope this book will have provided you with some new ways to think about it. But ultimately, optimism is irrational. You have to find the version you believe in for it to work for you. And I do hope it works for you.

The Consensus Cosmogony

Stories shape our ambitions, have the power to become real. Asimov's *Foundation* series inspired many real people to do real things: Newt Gingrich to reshape American politics; Martin Seligman to predict presidential races; and Elon Musk to found SpaceX, the company he created with the aim of making space travel cheap enough for humanity to become a multi-planetary species. "The lesson I drew from it is you should try to take the set of actions that are likely to prolong civilization, minimize the probability of a dark age and reduce the length of a dark age if there is one," Musk told *Rolling Stone* in 2017. The following year, he launched a Tesla Roadster into interplanetary space: it carried a quartz disc etched with the text of the Foundation trilogy. Ugh. So tacky.

But Musk and Asimov were themselves acting out a bigger story: a "consensus cosmogony" that had begun with Olaf Stapledon's 1937 epic *Star Maker* and then become the standard story of how humanity can reach the stars. In this picture, humanity first colonises our solar system; then reaches the stars; a galactic empire rises, then falls; a new empire arises; and eventually humanity transcends its earthly origins to become as gods. Asimov wrote the definitive version of the galactic empire parts; *Star Trek* later reinforced much the same narrative with the more democratic "Federation of Planets."

You don't have to believe this is what's literally going to happen: there's not going to be a galactic empire any time soon. But it is this cosmogony that underpins, say, Musk's conviction that terraforming Mars is our next logical step as a species (as opposed to, say, renewing our *own* sickly planet). Some of the world's most powerful—and optimistic—men have bought into this idea, with all its trappings of exploration, liberty and limitless expansion, and they are willing to throw billions of dollars at it.

I'd like to believe in this objective, this reach for the stars, but I don't think our current trajectory is towards *Star Trek*, or even the Galactic Empire. I think it's a bid to create Planet B—a refuge for a select few from the troubles of the Earth. That approach is fundamentally mistaken. Trying to leave the planet is a profoundly optimistic enterprise, and there are limits to any single individual's optimism, no matter how rich or powerful they are.

I think of Karl Popper, the philosopher of science and defender of liberal democracy, who, following in the train of Immanuel Kant, wrote: "Optimism is a duty. The future is open. It is not predetermined. No one can predict it, except by chance. We all contribute to determining it by what we do. We are all equally responsible for its success." There's a bit more to this statement than meets the eye. Not only is there the idea of practical reason—that optimism is about *making* the future brighter—but there's also the idea of optimism as a shared duty.

Why shared? Because when it comes to global challenges like climate change, or nuclear weapons, or artificial intelligence, no one person can do enough. There's a collective action problem: we need to be sure that if we do our part, everyone else will, too. But since we can't ever know that, we need to do our part without expectation of reward. That's the duty part. And we need to be optimistic about *each other*. We have to trust that we are more cooperative than competitive, more generous than selfish, more willing to struggle and challenge and overcome than to sit back, wait and watch.

That's why Shackleton said optimism was true moral courage: the courage to believe and act in the expectation—the "unrealistic" expectation—that others will, too. It's that sense of duty—shared duty—that got Shackleton and his men to safety in their tiny lifeboats. If we ever cross the far vaster and colder ocean of space between us and our neighbouring planet, in our tiny metal lifeboats, it'll be that same sense of duty that we need. And if we forget those escapist fantasies and stick to salvaging the planet we live on, it'll still be that sense of duty we'll need. Optimism is powerful, but we can't do it alone. No one's belief is strong enough to hold the world alone. We have to believe in better, together.

Forty Degrees and Rising

The cardboard is up over the windows of my kitchen in London, and it is . . . well, not scorching. I've been working in a room that is mercifully but guilt-inducingly cool, thanks to an air-conditioning unit humming loudly in the corner. It's a recursion of the problem: the planet is warming up, so I'm running a machine that consumes energy, which . . . warms the planet up. That my energy supplier guarantees me "green" electricity doesn't assuage my conscience: right now, more consumption still ultimately means more emissions.

This is not the hottest I've ever been. I've been on *holiday* to places where it was this hot. I've been to Death Valley; I've been to Egypt; I've been to Uluru; I've been to India. When I visited Delhi as a child, I wondered how people could live in these temperatures: it was a mercy when my relatives finally got air-conditioning, which I begged them to run, despite its cost, at every opportunity. When the power went out, as it did pretty much every day, we ran a diesel generator instead—just as millions still do. Dirty fuel to solve a problem caused by dirty fuel.

I wonder what it's like outside, so I open the door, and the gust of hot air is like the hair dryer blast that used to hit me when I stepped off the plane at Dum Dum Airport in Calcutta. And the smells are the same, the smell of dehydration and melting tarmac. It's not the smell of an English summer; it's the smell of an Indian city. I realise, decades later, that it wasn't the smell of India: it was the smell of heat. And the smell of the future.

The garden is yellow and brown, the leaves are withered, the grass crunches underfoot. There are no birds singing, no insects buzzing in the air. It's only been a couple of weeks of unseasonal heat, but some of the plants—some of the trees, even—probably won't make it. Perhaps they would if I gave them some water, but it seems vaguely pointless: if they survive this heatwave, they probably won't survive the next one, or the one after that. And hosepipes are banned anyway.

The advice is to replant with drought-resistant plants: spike-leafed, scrubby things that cling to the ground. It doesn't appeal, but it's not as if I am the gardener, anyway: my wife does that. My second, extraordinary wife, the woman who found her way to me, as I did to her, because we were both looking for the best of all possible worlds. This is the home we made, this is the garden we've cultivated. But that is not enough in this implacable heat—blanketing the planet, melting its ice, bleaching its coral, burning its forests. That will take something more. Something grander.

Better to Have Been

My phone is full of horrors. Day after day, hour after hour, it pipes in news of torments around the world: war, hunger, violence, abuse. I used to be able to ignore stories like these. I knew terrible things happened in the world. But I could tell myself they were aberrations, as indeed they are, statistically. Statistically, violence has never been lower. Statistically, we have never been healthier. Statistically, we have never been richer or wiser or happier. But the question still arises, unbidden: How can *this* possibly be the best of all possible worlds?

And yet it is, because this is the only world that's possible. From the cosmic to the quotidian, this is the world we live in. This is the world we have. It is the *only* world we have. Those other worlds, those other bubbles and branches of cosmic possibility: they have, for all intents and purposes, no physical existence. We can't travel to them; we can't touch them. They exist only in our minds, in the space of our imaginations.

The cosmic bubbles represent the spaces where our intellects weigh and assess the laws of the universe in our continuing bid to understand how its materials and forces work. Gravity, electricity, magnetism; logic, mathematics, computing; chemistry, biology, psychology. For as long as we keep testing our understanding of the universe—asking what we can expect of it, and what it will permit—then we will continue to advance our station, in this world and in the ones that have yet to come.

And the quantum branches are our decisions and actions, the paths we take and the accidents we suffer. For every choice we make, for every direction we travel, there are so many others we can't or won't take. We can think about the paths not taken, and the paths we have yet to take. We should use our imagination, our creativity, our craft, to tell the stories we want to come true: the stories of the world that we wish existed, that we wish we lived in and that our children, and their children will live in.

And for both the bubbles and the branches, for our grand designs and our gardens, our intellects and imaginations, we need our instinct and our intuition. We need the faith that leads to action, the belief that things can and will get better. That's the faith that brought me back home, the faith that won me back my life. It's the faith that leads us *all* to action, that can bring us *all* home, whether by reshaping the world or cultivating it.

Optimism in the face of the unknown future, willingness to accept and confront uncertainty: that's true moral courage, not the passivity of "realism." Change—progress—is possible. We owe it to ourselves, to those we have brought into the world, and to those who have yet to be born, to reach into the unknown in the hope of finding answers there. That's our shared duty: if we don't believe we can solve our problems, we *won't* solve them.

We're born optimistic. Some of us stay that way. If we get lost, we should try to make our way back.

The card is black, the card is white; the glass is empty, the glass is full. The world is light; the world is dark.

It's all how you look at it.

Choose the bright side.

Author's Note on Sources

If you'd like to know more about the subjects covered in this book you'll find a full list of sources at alternity.com.

There, you'll also find recommendations for further reading, footnotes and anecdotes that wouldn't fit into the book, and more of my writing on related ideas; you can also sign up for regular updates.

Thanks for reading!

Acknowledgments

The production of any book is an exercise in optimism: at its outset, all concerned are embarking on a long and winding odyssey whose destination is inevitably uncertain; each step forward is taken only because an expanding succession of people believe it's worth developing an idea into a manuscript and then into the object you're now holding.

This book is no exception. I became interested in optimism and in many-worlds thinking years ago, and these themes were developed through many conversations over many years. I neglected to record most of these conversations, because I wasn't working towards a book; but a few pivotal ones that stand out in my memory were with Patrick Bergel, Richard Fisher, Matt Webb and Scott David. My apologies to the many others who tolerated my ramblings, and contributed their insights, over decades past.

Nonetheless, I wasn't convinced anyone would be interested in my ruminations. That changed because of the persistence of my indefatigably positive agent Max Edwards, whose self-fulfilling prophecy was the proposal which eventually turned into this book. Thank you, Max, for overcoming my protestations; and thanks to the team at ACM UK—Vanessa Kerr, Gus Brown, Checkie Hamilton and Tom Lloyd-Williams—for your support at every step of the way. It's easier to be optimistic when you're confident you have the expertise you need on your side.

The proposal became a book thanks to my editors Helena Gonda and Rick Horgan, to whom I am eternally grateful for their demonstration of faith—especially after I fell foul of a bad case of the planning fallacy early on. (Physician, heal thyself!) The consecutive drafts of the manuscript explored many—a great many—possibilities when it came to both form and content; their thoughtful advice was critical in pruning them back to the one that remains.

What remains is undoubtedly not a perfect book, but it is the best of all the possible books I could have written, given the constraints of time, word count and human fallibility. I could not have achieved this optimality were it not for the contributions of Sophie Guimaraes, whose detailed suggestions included the single most difficult but valuable edit to the final book; Eugenie Todd, for her punctilious copyeditor's eye; and Laura Wise, for her patient diligence in production. Whatever imperfections remain are, of course, my responsibility and mine alone.

That's the end of the story of how this book was made. Its beginning, as you'll have read, lies in a incident, and a realisation, that changed my life. My reinvention as an optimist wouldn't and couldn't have happened without my friends Tracey, Matthew and Esme Sugden; Mathew Hyde; Clare Gilbert; and Kevin Hicks: I am more grateful than I can possibly say for your support at that time. My identification as a writer wouldn't have happened without the Million Monkeys: David Gullen, Tom Pollock, Gaie Sebold, Helen Callaghan and Sarah Ellender. And this book would never have been written without the energetic and enduring encouragement of Sally Adee and Melanie Garrett.

Finally, *none* of this would have been possible were it not for the love and patience of my family: my parents, my children and most of all my wife, Lucia, whose belief in me and this book has verged on the maritorious. Thank you for giving me the time and space to write it—and to be an optimist, each and every day.

Index